Theory of Graded-Bandgap Thin-Film Solar Cells

Synthesis Lectures on Electromagnetics

Editor
Akhlesh Lakhtakia, *The Pennsylvania State University*

Theory of Graded-Bandgap Thin-Film Solar Cells
Faiz Ahmad, Akhlesh Lakhtakia, and Peter B. Monk
2021

Spoof Plasmons
Tatjana Gric
2020

The Transfer-Matrix Method in Electromagnetics and Optics
Tom G. Mackay and Akhlesh Lakhtakia
2020

Theory of Graded-Bandgap Thin-Film Solar Cells

Faiz Ahmad, Akhlesh Lakhtakia, and Peter B. Monk

ISBN: 978-3-031-00896-2   paperback
ISBN: 978-3-031-02024-7   ebook
ISBN: 978-3-031-00139-0   hardcover

DOI 10.1007/978-3-031-00139-0

A Publication in the Springer series
SYNTHESIS LECTURES ON ELECTROMAGNETICS

Lecture **3**
Series Editor: Akhlesh Lakhtakia, *The Pennsylvania State University*
Series ISSN
ISSN pending.

Theory of Graded-Bandgap Thin-Film Solar Cells

Faiz Ahmad
COMSATS University Islamabad

Akhlesh Lakhtakia
The Pennsylvania State University

Peter B. Monk
University of Delaware

SYNTHESIS LECTURES ON ELECTROMAGNETICS 3

ABSTRACT

Thin-film solar cells are cheap and easy to manufacture but require improvements as their efficiencies are low compared to that of the commercially dominant crystalline-silicon solar cells. An optoelectronic model is formulated and implemented along with the differential evolution algorithm to assess the efficacy of grading the bandgap of the CIGS, CZTSSe, and AlGaAs photon-absorbing layer for optimizing the power-conversion efficiency of thin-film CIGS, CZTSSe, and AlGaAs solar cells, respectively, in the two-terminal single-junction format. Each thin-film solar cell is modeled as a photonic device as well as an electronic device. Solar cells with two (or more) photon-absorbing layers can also be handled using the optolelectronic model, whose results will stimulate experimental techniques for bandgap grading to enable ubiquitous small-scale harnessing of solar energy.

KEYWORDS

AlGaAs, bandgap grading, CIGS, CZTSSe, differential evolution algorithm, double-absorber solar cell, optoelectronic optimization, thin-film solar cell

Dedicated to the future availability of
inexpensive energy on a pollution-free planet

Contents

Preface

The poor remain poor because they cannot afford the energy needed for economic advancement. The poor also remain poor because they suffer disproportionately from pollution engendered by affluence. Their low energy use does not insulate them from the high greenhouse-gas emissions not of their making. This is all the more infelicitous because the annual global energy consumption grew in 2018 by 2.9% (double the 1.5% decadal average) and will continue to grow for several decades.

Since fossil fuels pollute the planet and warm up the atmosphere, fissile materials require high security, and hydropower dams wreck ecologies, we must turn toward green sources of energy. Photovoltaic solar cells are perhaps the most widely accessible green-energy sources, but contributed less than 1% of all primary energy consumed worldwide in 2018. Although ongoing improvements in energy-storage technology are promoting the construction of solar parks, that will not suffice to meet the needs of many of the poor who live in infrastructure-deficient locales. Moreover, solar parks remove land from farming and the transportation of electricity to the point of use adds transmission losses.

Decentralized and even in-device photovoltaic generation of electricity is needed in addition to solar parks. Future photovoltaic electricity generators (PEGs) must be able to convert ambient light, whether solar or from lighting fixtures, into electricity. PEGs must be thin, lightweight, rollable for ease in fabrication as well as transport, and deployable as wallpapers and fabrics. Easily replaceable patches of milliwatt PEGs could power small electrical devices.

The minimum efficiency for widespread adoption of a new solar cell is seen as 30% with the two-terminal single-junction format, which is the least wasteful format for harnessing solar energy at any wattage level. Crystalline-silicon (c-Si) solar cells are no more than 27.6% efficient. Even tandem c-Si/perovskite solar cells do not exceed 29.1% in efficiency, and the perovskite component fails within a year. Perovskite, organic, and quantum-dot solar cells have not delivered in excess of 25.2% efficiency, and their useful lifetime is brief. Thin-film solar cells can fulfill the requirements for ubiquitous electricity generation at the low wattage level, but these devices have not crossed 23% in efficiency.

Driven by the aim of successfully overcoming the perilous climate emergency, we have optoelectronically modeled thin-film solar cells with graded-bandgap photon-absorbing layers and found that bandgap grading enhances the efficiency. The bandgap of a compound semiconductor depends on its chemical composition. As bandgap grading through compositional grading affects local photonic and electronic properties, optimal bandgap grading requires detailed understanding of the solar cell as a photonic device as well as an electronic device.

This monograph presents our optoelectronic model and its implementation for thin-film solar cells with a graded-bandgap photon-absorbing layer of CIGS, CZTSSe, or AlGaAs. Furthermore, the use of two graded-bandgap photon-absorbing layers (instead of one) in one thin-film solar cell is shown to cross the 30% threshold. These results suggest that bandgap grading is a promising approach for achieving higher efficiency in thin-film solar cells. It is our great hope that this approach will be tried experimentally.

Beyond the particular studies we present here, we intend this monograph to help improve the state-of-the-art in simulation and optimization of thin-film solar cells by presenting details of the models and algorithms in a consistent and, we hope, clear way. The combination of mathematical analysis, computational modeling, and engineering is a potent approach to any major design problem, and one that has been both exciting to undertake and scientifically productive.

Faiz Ahmad, Akhlesh Lakhtakia, and Peter B. Monk
August 2021

Acknowledgments

Faiz Ahmad owes a special thanks to his wife, Faiza Khan, for all her love and support during his doctoral studies and subsequently.

Akhlesh Lakhtakia is grateful for conversations on sustainability with his colleagues Drs. Darryl Farber and Torben A. Lenau. Stimulating interactions during Fall 2020 with students Nitis Chantarawong, Illif Grady, Michael Greve, Chengzhi Li, Aaryan Oberoi, Justin Pierce, Madison Reddie, and Kanembe Shanachilubwa will always be fondly remembered by him. He also acknowledges the Charles Godfrey Binder Endowment at Penn State for ongoing support of his research activities.

Peter B. Monk thanks the US Air Force Office of Scientific Research and Dr. Arje Nachman, the program officer who funded his early research on Maxwell's equations that made the work reported in this monograph possible. He is also grateful to Dr. Muhammad Faryad (Lahore University of Management Sciences) for his patience when explaining solar-cell concepts during the early stages of collaborative work with Penn State on optical modeling of solar cells.

All three of us are grateful to Drs. Tom H. Anderson, Benjamin J. Civiletti, and Yangwen Zhang for wonderful collaboration facilitated by the US National Science Foundation under Grant Nos. DMS-1619901, DMS-1619904, DMS-2011996, and DMS-2011603. Finally, we thank the staff of Morgan & Claypool for splendid cooperation in producing this book.

Faiz Ahmad, Akhlesh Lakhtakia, and Peter B. Monk
August 2021

Acronyms and Symbols

ACRONYMS AND CHEMICAL SYMBOLS

Ag	Silver	FSP	Front-surface passivation
Al	Aluminum	Ga	Gallium
AlGaAs	Aluminum-gallium arsenide	GaAs	Gallium arsenide
AlInP	Aluminum-indium phosphide	GaInP	Gallium-indium phosphide
Al_2O_3	Aluminum oxide	Ge	Germanium
AM1.5G	Air mass 1.5 global	MgF_2	Magnesium fluoride
As	Arsenic	Mo	Molybdenum
a-Si:H	Hydrogenated amorphous silicon	od-ZnO	Oxygen-deficient zinc oxide
Au	Gold	Pd	Palladium
AZO	Aluminum-doped zinc oxide	RCWA	Rigorous coupled-wave approach
BSP	Back-surface passivation	S	Sulfur
CdS	Cadmium sulfide	Se	Selenium
CdTe	Cadmium telluride	SRH	Shockley–Read–Hall
CIGS	Copper indium-gallium diselenide	Sn	Tin
c-Si	Crystalline silicon	TCO	Transparent conducting oxide
Cu	Copper	Zn	Zinc
CZTSSe	Copper zinc tin sulfide-selenide	ZnS	Zinc sulfide
DEA	Differential evolution algorithm		

PRINCIPAL SCALARS

Symbol	Units	Description
A		Bandgap perturbation amplitude parameter
$\bar{a}_s, \bar{a}_p$		Scalar amplitudes of s- and p-polarized components of incident plane wave
$a_s^{(m)}, a_p^{(m)}$		Scalar amplitudes of m-th order s- and p-polarized components of incident field
A_s, A_p		Linear absorptances for s- and p-polarized illumination
c_0	$= 3 \times 10^8$ m s^{-1}	Speed of light in free space
$\tilde{C}$		Crossover fraction for DEA
C_n	cm^6 s^{-1}	Auger recombination coefficient for electrons
C_p	cm^6 s^{-1}	Auger recombination coefficient for holes
E_0	V m^{-1}	Amplitude of the incident electric field
E_0	eV	Arbitrary reference energy
E_c	eV	Conduction band-edge energy
E_F	eV	Fermi level
E_{F_n}	eV	Electron quasi-Fermi level
E_{F_p}	eV	Hole quasi-Fermi level
E_g	eV	Bandgap energy
E_i	eV	Intrinsic energy of charge carriers
E_T	eV	Trap energy level for SRH recombination
E_v	eV	Valence band-edge energy
E_{vac}	eV	Vacuum energy
F		Probability of occupation of an energy state
F_{MB}		Maxwell–Boltzmann distribution function
$\tilde{F}$		Stepsize for DEA
FF		Fill factor
G	cm^{-3} s^{-1}	Electron-hole-pair generation rate
$\hbar$	$= 1.054 \times 10^{-34}$ J s	Reduced Planck constant
i		$\sqrt{-1}$
J_{dev}	mA cm^{-2}	Total device current density

Symbol	Units	Description
J_n	mA cm^{-2}	Electron current density
J_p	mA cm^{-2}	Hole current density
J_{sc}	mA cm^{-2}	Short-circuit current density
J_{sc}^{opt}	mA cm^{-2}	Optical short-circuit current density
K		Bandgap perturbation period number
k_0		Wavenumber in free space
k_B	$= 1.3806 \times 10^{-23}$ J K^{-1}	Boltzmann constant
L_{BSP}	m	Thickness of back-surface passivation layer
L_{FSP}	m	Thickness of front-surface passivation layer
L_g	m	Corrugation height
L_m	m	Thickness of metal layer
L_s	m	Thickness of main photon-absorbing layer
L_t	m	Total thickness of solar cell
L_{TCO}	m	Thickness of TCO layer
L_w	m	Thickness of window layer
L_x	m	Period of periodically corrugated back reflector
M_t		Maximum index for summation in RCWA (for computational tractability)
n	cm^{-3}	Electron density
$\tilde{N}$		Number of control parameters for optimization
N_0	cm^{-3}	Baseline number density
n_1	cm^{-3}	Electron density at defect/trap energy level for SRH recombination
N_A	cm^{-3}	Acceptor density/concentration
N_c	cm^{-3}	Density of states in conduction band
N_D	cm^{-3}	Donor density/concentration
N_d	cm^{-3}	$= N_D - N_A$
N_f	cm^{-3}	Fixed-charge defect/trap density
n_i	cm^{-3}	Intrinsic charge-carrier density
N_P		Number of points selected for optimization
N_v	cm^{-3}	Density of states in valence band
P	W m^{-2}	Power density obtained from solar cell

Symbol	Units	Description
$\mathfrak{P}_\ell$		ℓ-th population for optimization
p	cm^{-3}	Hole density
p_1	cm^{-3}	Hole density at trap energy level for SRH recombination
P_{in}	W m^{-2}	Incident power density obtained by integrating $S(\lambda_0)$ over the solar spectrum
P_{max}	W m^{-2}	Maximum power density obtained from solar cell
Q	J m^{-3}	Monochromatic optical energy per unit volume
q_e	$= 1.602 \times 10^{-19}$ C	Elementary charge
R	cm^{-3} s^{-1}	Electron-hole recombination rate
$\mathbb{R}$		Set of real numbers
R_{Aug}	cm^{-3} s^{-1}	Auger recombination rate
R_B	cm^3 s^{-1}	Radiative recombination coefficient
R_{rad}	cm^{-3} s^{-1}	Radiative recombination rate
R_{SRH}	cm^{-3} s^{-1}	SRH recombination rate
$r_s^{(m)}, r_p^{(m)}$		Scalar amplitudes of m-th order s- and p-polarized components of reflected field
$r_{ss}^{(m)}, r_{pp}^{(m)}$		Reflection coefficients of m-th order
$R_{ss}^{(m)}, R_{pp}^{(m)}$		Reflectances of m-th order
S	W m^{-2} nm^{-1}	Solar irradiance
$\mathbb{S}$		Set of all possible choices of optimization parameters
T	K	Absolute temperature
$t_s^{(m)}, t_p^{(m)}$		Scalar amplitudes of m-th order s- and p-polarized components of transmitted field
$t_{ss}^{(m)}, t_{pp}^{(m)}$		Transmission coefficients of m-th order
$T_{ss}^{(m)}, T_{pp}^{(m)}$		Transmittances of m-th order
V_{ext}	V	External bias voltage
V_{oc}	V	Open-circuit voltage
V_{th}	V	Thermal voltage
v_{th}	cm s^{-1}	Mean thermal speed
$v_{th,n}$	cm s^{-1}	Mean thermal speed of electrons
$v_{th,p}$	cm s^{-1}	Mean thermal speed of holes
$\tilde{\mathbf{x}}$		Control vector for optimization
α		Bandgap perturbation shape parameter

Symbol	Units	Description
ε_0	$= 8.854 \times 10^{-12}$ F m^{-1}	Permittivity of free space
ε	F m^{-1}	Permittivity
$\varepsilon^{(m)}$	F m^{-1}	m-th order coefficient of Fourier series of permittivity ε
ε_d	F m^{-1}	Permittivity of Al$_2$O$_3$
ε_{dc}		DC relative permittivity
ε_g	F m^{-1}	Permittivity in grating region
ε_m	F m^{-1}	Permittivity of metal
η		Solar-cell efficiency
η_0	$= 120\pi$ Ω	Intrinsic impedance of free space
λ_0	m	Free-space wavelength
μ_0	$= 1.2571 \times 10^{-6}$ H m^{-1}	Free-space permeability
μ_n	cm^2 V^{-1} s^{-1}	Electron mobility
μ_p	cm^2 V^{-1} s^{-1}	Hole mobility
ν		Bandgap perturbation phase parameter
ω	rad-Hz	Angular frequency
ϕ	V	DC electric potential
ϕ_n	V	built-in potential for electrons
ϕ_p	V	built-in potential for holes
σ_n	cm^2	Capture cross section for electrons
σ_p	cm^2	Capture cross section for holes
τ_n	s	Minority carrier lifetime of electrons
τ_p	s	Minority carrier lifetime of holes
$\xi : (1 - \xi)$		Proportion of gallium relative to that of indium in CIGS photon-absorbing layer
		Proportion of sulfur relative to that of selenium in CZTSSe photon-absorbing layer
		Proportion of aluminum relative to that of gallium in AlGaAs photon-absorbing layer
$\bar{\xi} : (1 - \bar{\xi})$		Proportion of aluminum relative to that of gallium in AlGaAs window layer
ζ		Duty cycle of corrugation of periodically corrugated back reflector

PRINCIPAL VECTORS AND MATRIXES

Symbol	Description
$\breve{\mathbf{A}}$	Column vector of phasor amplitudes of the incident field
$\mathbf{B}$	Magnetic field phasor
$\mathbf{D}$	Electric displacement field phasor
$\mathbf{E}$	Electric field phasor
$\mathbf{e}$	Auxiliary electric field phasor
$\breve{\mathbf{f}}$	Column vector containing x- and y-directed components of $\mathbf{e}$ and $\mathbf{h}$
$\mathbf{H}$	Magnetic induction field phasor
$\mathbf{h}$	Auxiliary magnetic induction field phasor
$\breve{\mathbf{I}}, \breve{\mathbf{1}}$	Identity matrix
$\breve{\mathbf{K}}$	Fourier-wavenumber matrix
$\breve{\mathbf{O}}, \breve{\mathbf{0}}$	Null matrix
$\breve{\mathbf{P}}$	Kernel matrix in matrix ordinary differential equation
$\breve{\mathbf{R}}$	Column vector of phasor amplitudes of the reflected field
$\mathbf{r} \equiv (x, y, z)$	Position vector
$\breve{\mathbf{T}}$	Column vector of phasor amplitudes of the transmitted field
$\hat{\mathbf{u}}_x$	Unit vector in x direction
$\hat{\mathbf{u}}_y$	Unit vector in y direction
$\hat{\mathbf{u}}_z$	Unit vector in z direction
$\breve{\mathbf{Z}}$	Auxiliary transmission matrix
$\breve{\varepsilon}$	Toeplitz matrix containing Fourier coefficients of permittivity

CHAPTER 1

Introduction

1.1 ENERGY AND ENVIRONMENT

The global energy demand is continuously rising with the growing population and improving living standards. The need for electrical power generation is expected to reach 23.5 terawatt (TW) in the year 2040, an increase of 27% from the current requirement of 18.5 TW [1]. This estimated increase is mainly due to the expected population increase, with 1.7 billion people to be added to the developing economies, and the economic growth of developing countries in Asia, especially China and India.

In the year 2018, energy generated using commercially traded fuels including biofuels, nuclear plants, hydroelectric dams, geothermal sources, tides, winds, and sunlight for consumption amounted to 13864.9 million tons oil equivalent (MTOE), of which all renewable sources contributed 561.3 MTOE (i.e., just 4%). Renewable sources include biofuels, tides, winds, and sunlight. The share of photovoltaic solar cells was about 112 MTOE (i.e., 0.8%). Fossil fuels such as oil, coal, and natural gas contributed 11743.6 MTOE (i.e., 84.7%) [2].

Three major concerns have emerged from the large-scale practice of burning carbonaceous fuels. Climate emergency is arguably the most significant of these three concerns. A direct consequence of adding CO_2, methane, and other gases to the atmosphere by burning these fuels is the greenhouse effect. Without cutting their consumption, global warming is likely to exceed 2°C above pre-industrialized levels by the year 2050. This would have a considerable impact on the landscape and sea levels. Effects of global climate change can already be felt in terms of polar ice melting, weather pattern changes, more heat waves, and more intense hurricanes [3].

Energy security is the second primary concern, as the supply of fossil fuels remains volatile. Terrestrial fossil-fuel resources are limited and shrinking with every passing day. It is projected that these resources will get depleted in the next hundred years or so [4]. Still, approximately 12% of the world's population has no access to electricity [1]. Therefore, it will be challenging to fulfill the ever-increasing energy demand with fossil fuels only.

The third concern is the degradation of air quality due to the burning of fossil fuels. With the increased release of poisonous gases such as sulfur dioxide and nitrogen oxides in the atmosphere, the prevalence of diseases such as burning lung tissue, asthma, bronchitis, pulmonary inflammation, and chronic respiratory illness is increasing [4]. Air pollution continues to result in millions of premature deaths each year [5].

1.2   ECO-RESPONSIBLE SOURCES OF ENERGY

The world needs more attention toward less polluting and renewable energy sources to fulfill the future energy demand as well as to tackle the problems of climate emergency and environmental degradation. The chief eco-responsible energy sources are hydroelectric dams, wind, biomass, geothermal wells, and the Sun [1]. Hydropower- and wind-based technologies need favorable locations to harness the energy, but these locations are limited. Also, these technologies are grid-based, and transportation of energy to remote areas is costly [6]. Geothermal energy is very location-specific, and most resources are not competitive due to hefty costs associated with both geothermal power plants and geothermal heating/cooling systems [7]. Biomass technology is based on biomaterials, but it puts undue pressure on agricultural activities [8].

Nuclear energy is another carbon-free source of energy. It can be an essential part of the energy mix necessary to meet future electricity demand. Currently, nuclear plants contribute 4.4% of the global energy consumption [2]. But, nuclear energy can only be a short-term solution because nuclear safety is a gigantic security problem [9].

Yet another promising carbon-free energy source is nuclear fusion [10]. The fusion fuel cycle is sustainable and nearly inexhaustible due to the abundant availability of deuterium from seawater and tritium from breeding reactors. The major challenges for harnessing fusion power are plasma confinement and the maintenance of a stable plasma-material interface. As no design has produced more energy than the input energy so far [11], this technology remains futuristic.

Solar energy falling on the Earth in one hour is more than the global energy consumption in one year [12]. With oil price increases and reduction of cost of photovoltaic solar cells [1], solar energy has emerged as a significant force to control global climate change and environmental degradation, and it can also help to tackle the ever-increasing energy demand. Among all non-polluting sources of energy, the availability of sunlight in most parts of the world makes it the most viable source of energy; plus, the Sun is an almost a never-ending source of energy. Equally importantly, the off-grid generation of solar energy has no transmission losses and can supply electricity in remote regions.

With all its benefits there are a few disadvantages of solar energy. Solar-cell modules can function only when it is sunny. Hence, batteries are needed for energy storage for off-time use [13], which adds cost. Fortunately, battery performance has been increasing rapidly [14]. Also, solar-cell modules used to harvest solar energy use valuable land that can be utilized otherwise for agriculture. Finally, one must consider pollution arising from the fabrication of solar cells and solar-cell modules as well as from their post-use disposal in carefully conducted life-cycle audits [15].

1.3   PHOTOVOLTAIC SOLAR CELLS

Photovoltaic solar cells, simply known as solar cells, directly convert sunlight into electricity via the photoelectric effect. The first solar cell, demonstrated at Bell Labs in 1954, was made of crystalline silicon (c-Si) and had 6% power-conversion efficiency [16]. Solar cell-technology made a significant advancement in the last decade. The current record efficiency for single-junction c-Si solar-cell modules is 26.7% [17]. Advanced solar-energy modules incorporating multi-junction solar cells illuminated by concentrated sunlight are much as 46% efficient [17, 18]. However, the efficiencies of commercially available solar cells (15–20%) are well below the laboratory solar-cell-module efficiency due to factors such as degradation of material properties and increased electron-hole-pair recombination rate upon prolonged solar illumination [19]. A chart of solar cell efficiencies provided by the National Renewable Energy Laboratory (NREL) [18] is shown in Fig. 1.1.

The installed solar-energy generation capacity grew from 0.65 GW in the year 2000 to 40.13 GW in 2010 to 586.4 GW in 2019 [20]. Despite the 15-fold increase during the last decade, the contribution of photovoltaic solar cells to global energy consumption did not exceed 1% in 2018 [2]. This contribution is expected to increase to 70% in 2050 [21]. Furthermore, the price of solar-cell modules has gone down considerably and is expected to go down further as the installed capacity increases [22].

Although solar cells in some countries almost match conventional sources in consumer price, a further cost reduction is required for solar cells to be fully competitive with conventional sources. Large-scale adoption of solar technology is hindered by factors such as low power-conversion efficiency, scarcity of materials, and low industrial-scale production.

The most critical factor for large-scale adoption of solar cells is the reduction of the unit cost. Two methods can reduce it. The first method is to increase the power-conversion efficiency of the solar cell. However, there is an upper bound on efficiency called the Shockley–Queisser limit [23]. This bound is based on the maximum radiative recombination of electrons and holes, but most materials in practice cannot reach that maximum. For a single-junction c-Si solar cell with a bandgap of 1.12 eV, the Shockley–Queisser limit is 33%, and the present-day record efficiency (26.7%) is in the ballpark [17].

The second method is to reduce the manufacturing cost while maintaining the power-conversion efficiency. The manufacturing cost can be reduced using less expensive materials, adopting less costly production methods, and reducing the thickness of the photon-absorbing layer in the solar cell [24]. Thin-film solar cells can provide all three of these features and thus have the potential to fulfill a significant fraction of future energy demand.

The other factors for economical photovoltaic technology are efficiency, energy-payback time, temporal stability (i.e., rate of module degradation due to use), and the availability of materials. The efficiency of solar-cell modules is an essential consideration for satellite applications and also at the system level. Higher efficiency reduces the number of modules as well as the area required for mounting the modules, the land needing preparation as well as area-dependent operation and maintenance costs [22]. Energy-payback time is the duration in which a solar-cell module can produce the amount of energy needed to fabricate the module itself. Short energy-payback time is desirable for economical solar power installation. The temporal stabil-

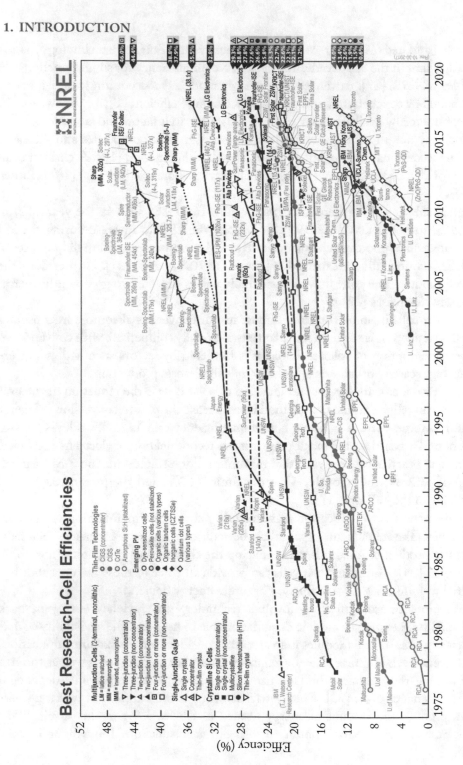

Figure 1.1: Efficiencies of the best solar cells reported by researchers [18].

ity of solar cells is another important factor. Some solar-cell technologies show great promise in a laboratory setting; however, their commercial feasibility is hindered by the degradation of the photon-absorbing materials over time, as exemplified by perovskites [25, 26] and hydrogenated amorphous silicon [19]. The scarcity and toxicity of materials used in solar cells are also important factors.

1.4   CRYSTALLINE-SILICON SOLAR CELLS

The global solar-cell market is dominated by c-Si solar cells due to high efficiency, the existence of a well-established semiconductor industry, and an abundance of silicon in the Earth's crust. The current market share of c-Si solar cells is 94%, and their maximum power-conversion efficiency is 26.7% [27].

There are two types of c-Si solar cells: monocrystalline and polycrystalline [28]. The fabrication of monocrystalline Si solar cells is done using the Czochralski process, in which a single crystal is drawn from a melt to form a long ingot, which is sliced into wafers of thickness 100–200 μm [28]. The commonly used fabrication method for polycrystalline solar cells is the Siemens process [28], which requires gasification of metallurgical-grade Si, distillation to reach the required purity, and finally deposition to obtain ultra-pure Si. Production of polycrystalline Si solar cells consumes less energy, but the efficiency (19.9%) is less compared to monocrystalline Si solar cells [17].

Commercial c-Si solar cells deliver close to the maximum achievable efficiency and further cost reduction seems difficult [29]. Even though their cost continues to drop, small-scale photovoltaic generation of energy must become ubiquitous for human progress to become truly unconstrained by energy economics. Thin-film solar cells are necessary for that to happen.

1.5   THIN-FILM SOLAR CELLS

Thin-film solar cells are a viable option compared to c-Si solar cells due to low cost and ease of manufacturing. The c-Si solar cells are made of highly pure silicon wafers that account for 50% of the total module cost. Thin-film solar cells are fabricated using physical vapor deposition and chemical vapor deposition [24]. As thin-film solar cells are made of materials having higher optical absorption than c-Si, the thickness of the photon-absorbing layer is on the order of a few hundred nanometers. In contrast, the typical thickness of the c-Si layer in a solar cell is a few hundred micrometers [30]. The energy-payback time for a thin-film solar cell is almost half compared to a c-Si solar cell [31]. Also, the flexibility of some thin-film solar cells makes them attractive for portable purposes such as camping and installation on nonplanar surfaces such as car roofs [32]. Another benefit of thin-film solar cells is the control over the composition of a material, i.e., the bandgap of material can be varied by varying the material composition.

Solar cells are named after the material used for the photon-absorbing layer, which is the major contributor to charge-carrier generation. Currently, thin-film solar cells containing photon-absorbing layers made of either $CuIn_{1-\xi}Ga_\xi Se_2$ (commonly referred as CIGS) or CdTe are commercially dominant [30]. Another type of commercial thin-film solar cell uses hydrogenated amorphous silicon (a-Si:H) for absorbing solar photons.

The maximum efficiency is 10.2% for a-Si:H, 22.6% for CIGS, and 21.0% for CdTe solar cells [17, 18]. These thin-film solar cells have 6% share of the global photovoltaic solar-cell market [33].

1.5.1 CHALLENGES

For widespread adoption, thin-film solar cells must be made of materials that are abundant on our planet, and the materials of choice must be those that can be extracted, processed, and discarded with low environmental impact. CIGS and CdTe are commercially dominant (terrestrial) thin-film solar cell technologies; however, there are serious concerns about the planetwide availability of indium and tellurium [34]. Hence, thickness reduction of the photo-absorbing layer is necessary to tackle the scarcity of indium in CIGS solar cells and of tellurium in CdTe solar cells. Furthermore, both indium and cadmium are toxic, leading to concerns about their environmental impact following disposal after use.

Another problem with thin-film solar cells is their low power-conversion efficiency compared to c-Si solar cells. $Cu_2ZnSn(S_\xi Se_{1-\xi})_4$ (commonly referred as CZTSSe) and a-Si:H solar cells contain Earth-abundant and non-toxic materials, but their efficiencies are only 12.6% and 10.2%, respectively. Both efficiencies are considerably lower than the approximately 22% efficiencies of the CdTe and CIGS solar cells [18]. The primary reason for the lower efficiency of CZTSSe solar cells is the lower lifetime of minority charge-carriers due to higher trap/defect density, which shortens their diffusion length, thereby limiting the collection of minority carriers deep in the CZTSSe layer [35]. The reasons for low power-conversion efficiency in a-Si:H solar cells are (i) the high electron-hole recombination rate and low charge-carrier diffusion lengths [36] and (ii) the phenomenon of light-induced degradation, also known as the Staebler–Wronski effect [19]. Therefore, the reduction of the thickness of the photon-absorbing layer is desirable for better collection of charge-carriers in both CZTSSe and a-Si:H thin-film solar cells.

GaAs thin-film solar cells have excellent efficiency (28.8%) and significantly lower weight-to-power ratio than c-Si solar cells [37]. GaAs solar-cell technology is the current market leader for extra-terrestrial applications, but it is prohibitively expensive for terrestrial applications [38]. Cost reduction by thinning the expensive GaAs photon-absorbing layer is necessary to make this technology viable for terrestrial applications.

The other thin-film solar-cell technologies, i.e., perovskites, organic, and dye-sensitized solar cells, have not penetrated the market yet in large volumes [17, 18]. However, perovskite/c-Si tandem solar cells appear to hold commercial promise [39, 40].

1.5.2 TRAPPING OF LIGHT

Reduction of the thickness of the photon-absorbing layer is therefore desirable to reduce the cost, improve charge-carrier collection, and overcome material scarcity. A thinner photon-absorbing layer will also improve the production throughput. However, a thinner layer will absorb fewer solar photons, which is inimical to power-conversion efficiency denoted by η. The lower absorption of solar photons would reduce both the optical short-circuit current density J_{sc}^{opt} and the open-circuit voltage V_{oc}.

Trapping of light is necessary to enhance the absorption of photons and, thus (hopefully), the power-conversion efficiency. The most straightforward light-trapping technique is to add a back reflector which effectively doubles the optical thickness of the photon-absorbing layer and enhances the charge-carrier-generation rate inside the solar cell. Several other techniques under investigation include the use of light-trapping nanostructures [41–45], back-surface passivation layers [46], anti-reflection coatings [47–49], textured front faces [50, 51], metallic periodically corrugated back reflectors [52–54], plasmonic particles [55], surface plasmonics [56–58], multiplasmonics [59–61], and waveguide-mode excitation [62–64]. These techniques are at best in the research stages.

1.5.3 BANDGAP GRADING OF PHOTON-ABSORBING LAYER

Another option is to use a compound semiconductor for the photon-absorbing layer and grade its bandgap by controlling the compositional ratio of elements in that semiconductor [65–68]. Bandgap grading can allow photon absorption over a wider frequency range. Also, bandgap grading will increase η by creating a drift electric field that will accelerate photogenerated holes toward the *p-n* junction inside the solar cell [67]. Linear grading of the bandgap has been experimentally shown to increase V_{oc} [69], which should assist in enhancing η; however, suboptimal grading can reduce the short-circuit current density J_{sc} to offset the increase in V_{oc}. Optimization is required for bandgap grading to maintain J_{sc} while enhancing V_{oc}.

1.6 OVERVIEW

It is commonplace in the optical literature on solar cells to optimize the optical response characteristics as quantified by the optical short-circuit current density J_{sc}^{opt} [44, 52, 53, 57, 59, 60, 64, 70–73]. But that is a deficient strategy as the maximization of J_{sc}^{opt} can reduce V_{oc} [74]. Hence, adequate modelling of a thin-film solar cell requires the coupling of (i) an optical submodel capable of capturing the absorption of photons and (ii) an electrical submodel capable of simulating the transport of charge carriers throughout the solar cell [75, 76]. Both submodels must be rapidly computable, accurate, and robust across the relevant parameter space for the optimization of solar-cell designs.

A comprehensive but not overly cumbersome model [77] is provided in this monograph for optoelectronic simulation of thin-film solar cells with graded-bandgap photon-absorbing layers, in order to tackle the material-scarcity and low-efficiency issues. Three different types of thin-film solar cells are addressed to exemplify the vast number of compound semiconductors that have been fabricated [78]: CIGS, CZTSSe, and $Al_\xi Ga_{1-\xi}As$ (commonly referred to as AlGaAs). The bandgap of a CIGS layer can be graded by dynamically controlling the compositional ratio of gallium and indium during fabrication. Similarly, the bandgaps of CZTSSe and AlGaAs can be varied by controlling the compositional ratios of sulfur and selenium, and aluminum and gallium, respectively. The optoelectronic model allows the incorporation of periodically corrugated back reflectors, back-surface passivation layers, and localized back-contacts. The model has been coupled with an optimization algorithm [79, 80] to maximize the power-conversion efficiency in relation to geometric and bandgap-grading parameters [81].

This monograph is organized as follows. In Chapter 2, the optics of semiconductors and the optical part of the coupled optoelectronic model are discussed. The frequency-domain Maxwell equations are presented first. The rigorous coupled-wave approach [82, 83], which is the engine for optical computations for thin-film solar cells with a periodically corrugated back reflector, is discussed next in detail. Expressions for the electron-hole-pair generation rate G and optical short-circuit density J_{sc}^{opt} are derived.

Chapter 3 is devoted to semiconductor physics required for the electronic part of the coupled optoelectronic model, with G serving as the input and η as the main output. The basics of semiconductors are discussed briefly. The semiconductor transport equations, also called the drift-diffusion equations, fundamental to solar-cell device physics are presented along with boundary conditions and mathematical expressions for recombination processes [75, 76]. The solution procedure for the one-dimensional drift-diffusion model by discretization using the hybridizable discontinuous Galerkin scheme [70, 84–87] is discussed briefly. The chapter closes with an introduction to the differential evolution algorithm [79, 80] used for optimization.

Chapters 4, 5, and 6 are devoted to the optoelectronic modeling and optimization of CIGS, CZTSSe, and AlGaAs thin-film solar cells, with numerical results presented on optoelectronic optimization for homogeneous and graded-bandgaps photon-absorbing layers of various thickness. An optimal CIGS⊕CZTSSe thin-film solar cell with two different photon-absorbing layers and possessing a power-conversion efficiency predicted to be in excess of 30% is presented at the end of Chapter 6.

1.7   BIBLIOGRAPHY

[1] International Energy Agency, *Data and Statistics* (accessed May 17, 2021). 1, 2

[2] British Petroleum, *BP Statistical Review of World Energy 2019*, 68th ed. (British Petroleum, London, UK, 2019). 1, 2, 3

[3] National Aeronautics and Space Administration, *Global Climate Change, Facts* (accessed May 17, 2021). 1

[4] Union of Concerned Scientists, *The Hidden Costs of Fossil Fuels* (accessed May 17, 2021). 1

[5] M. Roser, *Energy Poverty and Indoor Air Pollution* (accessed July 5, 2021). 1

[6] F. H. Saadi, N. S. Lewis, and E. W. McFarland, Relative costs of transporting electrical and chemical energy, *Energy and Environmental Science*, 11:469–475 (2018). 2

[7] Z. Hyder, *Geothermal Energy Pros and Cons* (accessed May 17, 2021). 2

[8] D. Pimentel, A. Marklein, M. A. Toth, M. N. Karpoff, G. S. Paul, R. McCormack, J. Kyriazis, and T. Krueger, Food versus biofuels: Environmental and economic costs, *Human Ecology*, 37:1–12 (2009). 2

[9] J. E. Doyle, Ed., *Nuclear Safeguards, Security, and Nonproliferation: Achieving Security with Technology and Policy*, 2nd ed. (Elsevier, Cambridge, MA, 2019). 2

[10] C. L. Smith and D. Ward, The path to fusion power, *Philosophical Transactions of Royal Society A*, 365:945–956 (2007). 2

[11] World Nuclear Association, *Nuclear Fusion Power* (accessed May 17, 2021). 2

[12] National Renewable Energy Laboratory, *Solar Energy Basics* (accessed May 17, 2021). 2

[13] P. Paniyil, V. Powar, R. Singh, B. Hennigan, P. Lule, M. Allison, J. Kimsey, A. Carambia, D. Patel, D. Carrillo, Z. Shriber, T. Bazer, J. Farnum, K. Jadhav, and D. Pumputis, Photovoltaics- and battery-based power network as sustainable source of electric power, *Energies*, 13:5048 (2020). 2

[14] K. Parkman, *Find the Best Solar Batteries* (accessed May 19, 2021). 2

[15] B. R. Bakshi, *Sustainable Engineering: Principles and Practice* (Cambridge University Press, Cambridge, UK, 2019). 2

[16] D. M. Chapin, C. S. Fuller, and G. L. Pearson, A new silicon $p - n$ junction photocell for converting solar radiation into electrical power, *Journal of Applied Physics*, 25:676–677 (1954). 3

[17] M. A. Green, Y. Hishikawa, E. D. Dunlop, D. H. Levi, J. Hohl-Ebinger, and A. W. Y. Ho-Baillie, Solar cell efficiency tables (version 51), *Progress in Photovoltaics: Research and Applications*, 26:3–12 (2018). 3, 5, 6

[18] National Renewable Energy Laboratory, *Best Research-Cell Efficiencies* (accessed May 17, 2021). 3, 4, 6

[19] A. Kołodziej, Staebler–Wronski effect in amorphous silicon and its alloys, *Opto-electronics Review*, 12:21–32 (2004). 3, 5, 6

[20] Our World in Data, *Installed Solar Energy Capacity* (accessed May 19, 2021). 3

[21] M. Ram, D. Bogdanov, A. Aghahosseini, A. Gulagi, S. A. Oyewo, M. Child, U. Caldera, K. Sadovskaia, J. Farfan, L. S. N. S. Barbosa, M. Fasihi, S. Khalili, B. Dalheimer, G. Gruber, T. Traber, F. De Caluwe, H.-J. Fell, and C. Breyer, *Global Energy System Based on 100% Renewable Energy: Power, Heat, Transport, and Desalination Sectors* (Lappeenranta University of Technology, Lappeenranta, Finland, and Energy Watch Group, Berlin, Germany, 2019). 3

[22] M. A. Green, Photovoltaic technology and visions for the future, *Progress in Energy*, 1:013001 (2019). 3

[23] W. Shockley and H. J. Queisser, Detailed balance limit of efficiency of $p - n$ junction solar cells, *Journal of Applied Physics*, 32:510 (1961). 3

[24] K. L. Chopra, P. D. Paulson, and V. Dutta, Thin-film solar cells: An overview, *Progress in Photovoltaics: Research and Applications*, 12:69–92 (2004). 3, 5

[25] J. Xiong, B. Yang, C. Cao, R. Wu, Y. Huang, J. Sun, J. Zhang, C. Liu, S. Tao, Y. Gao, and J. Yang, Interface degradation of perovskite solar cells and its modification using an annealing-free TiO_2 NPs layer, *Organic Electronics*, 30:30–35 (2016). 5

[26] J. Yang, B. D. Siempelkamp, D. Liu, and T. L. Kelly, Investigation of $CH_3NH_3PbI_3$ degradation rates and mechanisms in controlled humidity environments using in situ techniques, *ACS Nano*, 9:1955–1963 (2015). 5

[27] Verband Deutscher Maschinen- und Anlagenbau, *International Technology Roadmap for Photovoltaic* (accessed May 17 2021). 5

[28] S. M. Sze, *Semiconductor Devices: Physics and Technology*, 2nd ed. (Wiley, New York, NY, 2002). 5

[29] L. C. Andreani, A. Bozzola, P. Kowalczewski, M. Liscidini, and L. Redorici, Silicon solar cells: Toward the efficiency limits, *Advances in Physics: X*, 4:1548305 (2018). 5

[30] M. A. Green, Thin-film solar cells: Review of materials, technologies, and commercial status, *Journal of Materials Science: Materials in Electronics*, 18:15–19 (2007). 5, 6

[31] K. P. Bhandari, J. M. Collier, R. J. Ellingson, and D. S. Apul, Energy payback time (EPBT) and energy return on energy invested (EROI) of solar photovoltaic systems: A systematic review and meta-analysis, *Renewable and Sustainable Energy Reviews*, 47:133–141 (2015). 5

[32] Powerfilm Solar, *About Powerfilm* (accessed May 17, 2021). 5

[33] Fraunhofer Institute for Solar Energy Systems, *Photovoltaics Report* (accessed May 17, 2021). 6

[34] C. Candelise, M. Winskel, and R. Gross, Implications for CdTe and CIGS technologies production costs of indium and tellurium scarcity, *Progress in Photovoltaics: Research and Applications*, 20:816–831 (2012). 6

[35] T. Gokmen, O. Gunawan, and D. B. Mitzi, Minority carrier diffusion length extraction in $Cu_2ZnSn(Se,S)_4$ solar cells, *Journal of Applied Physics*, 114:114511 (2013). 6

[36] D. E. Carlson and C. R. Wronski, Amorphous silicon solar cell, *Applied Physics Letters*, 28:671–673 (1976). 6

[37] A. van Geelen, P. R. Hageman, G. J. Bauhuis, P. C. van Rijsingen, P. Schmidt, and L. J. Giling, Epitaxial lift-off GaAs solar cell from a reusable GaAs substrate, *Materials Science and Engineering B*, 45:162–171 (1997). 6

[38] K. A. W. Horowitz, T. Remo, B. Smith, and A. Ptak, Techno-economic analysis and cost reduction roadmap for III-V solar cells, *NREL Technical Report NREL/TP-6A20-72103* (2018). 6

[39] Y. Hu, L. Song, Y. Chen, and W. Huang, Two-terminal perovskites tandem solar cells: Recent advances and perspectives, *Solar RRL*, 3:1900080 (2019). 6

[40] M. Jošt, L. Kegelmann, L. Korte, and S. Albrecht, Monolithic perovskite tandem solar cells: A review of the present status and advanced characterization methods toward 30% efficiency, *Advanced Energy Materials*, 10:1904102 (2020). 6

[41] R. Singh, G. F. Alapatt, and A. Lakhtakia, Making solar cells a reality in every home: Opportunities and challenges for photovoltaic device design, *IEEE Journal of the Electron Devices Society*, 1:129–144 (2013). 7

[42] R. J. Martín-Palma and A. Lakhtakia, Progress on bioinspired, biomimetic, and bioreplication routes to harvest solar energy, *Applied Physics Reviews*, 4:021103 (2017). 7

[43] M. Schmid, Review on light management by nanostructures in chalcopyrite solar cells, *Semiconductor Science and Technology*, 32:043003 (2017). 7

[44] J. Goffard, C. Colin, F. Mollica, A. Cattoni, C. Sauvan, P. Lalanne, J.-F. Guillemoles, N. Naghavi, and S. Collin, Light trapping in ultrathin CIGS solar cells with nanostructured back mirrors, *IEEE Journal of Photovoltaics*, 7:1433–1441 (2017). 7

[45] C. van Lare, G. Yin, A. Polman, and M. Schmid, Light coupling and trapping in ultra-thin $Cu(In,Ga)Se_2$ solar cells using dielectric scattering patterns, *ACS Nano*, 9:9603–9613 (2015). 7

[46] B. Vermang, J. T. Wätjen, V. Fjällström, F. Rostvall, M. Edoff, R. Kotipalli, F. Henry, and D. Flandre, Employing Si solar cell technology to increase efficiency of ultra-thin $Cu(In,Ga)Se_2$ solar cells, *Progress in Photovoltaics: Research and Applications*, 22:1023–1029 (2014). 7

[47] İ. G. Kavakli and K. Kantarli, Single and double-layer antireflection coatings on silicon, *Turkish Journal of Physics*, 26:349–354 (2002). 7

[48] S. K. Dhungel, J. Yoo, K. Kim, S. Jung, S. Ghosh, and J. Yi, Double-layer antireflection coating of MgF_2/SiN_x for crystalline silicon solar cells, *Journal of the Korean Physical Society*, 49:885–889 (2006). 7

[49] S. A. Boden and D. M. Bagnall, Sunrise to sunset optimization of thin film antireflective coatings for encapsulated, planar silicon solar cells, *Progress in Photovoltaics: Research and Applications*, 17:241–252 (2009). 7

[50] W. H. Southwell, Pyramid-array surface-relief structures producing antireflection index matching on optical surfaces, *Journal of the Optical Society of America A*, 8:549–553 (1991). 7

[51] K. C. Sahoo, M.-K. Lin, E.-Y. Chang, T. B. Tinh, Y. Li, and J.-H. Huang, Silicon nitride nanopillars and nanocones formed by nickel nanoclusters and inductively coupled plasma etching for solar cell application, *Japanese Journal of Applied Physics*, 48:126508 (2009). 7

[52] P. Sheng, A. N. Bloch, and R. S. Stepleman, Wavelength-selective absorption enhancement in thin-film solar cells, *Applied Physics Letters*, 43:579–581 (1983). 7

[53] C. Heine and R. H. Morf, Submicrometer gratings for solar energy applications, *Applied Optics*, 34:2476–2482 (1995). 7

[54] M. Solano, M. Faryad, A. S. Hall, T. E. Mallouk, P. B. Monk, and A. Lakhtakia, Optimization of the absorption efficiency of an amorphous-silicon thin-film tandem solar cell backed by a metallic surface-relief grating, *Applied Optics*, 52:966–979 (2013). 7

M. Solano, M. Faryad, A. S. Hall, T. E. Mallouk, P. B. Monk, and A. Lakhtakia, Optimization of the absorption efficiency of an amorphous-silicon thin-film tandem solar cell backed by a metallic surface-relief grating, *Applied Optics*, 54:398–399 (2015) (erratum).

[55] Y. Zhang, B. Jia, Z. Ouyang, and M. Gu, Influence of rear located silver nanoparticle induced light losses on the light trapping of silicon wafer-based solar cells, *Journal of Applied Physics*, 116:124303 (2014). 7

[56] L. M. Anderson, Parallel-processing with surface plasmons. A new strategy for converting the broad solar spectrum, *Proc. of 16th IEEE Photovoltaic Specialists Conference*, 1:371–377, San Diego, CA, September 27–30, 1982. 7

[57] L. M. Anderson, Harnessing surface plasmons for solar energy conversion, *Proc. of SPIE*, 408:172–178 (1983). 7

[58] M. G. Deceglie, V. E. Ferry, A. P. Alivisatos, and H. A. Atwater, Design of nanostructured solar cells using coupled optical and electrical modeling, *Nano Letters*, 12:2894–2900 (2012). 7

[59] M. Faryad and A. Lakhtakia, Enhancement of light absorption efficiency of amorphous-silicon thin-film tandem solar cell due to multiple surface-plasmon-polariton waves in the near-infrared spectral regime, *Optical Engineering*, 52:087106 (2013). 7

M. Faryad and A. Lakhtakia, Enhancement of light absorption efficiency of amorphous-silicon thin-film tandem solar cell due to multiple surface-plasmon-polariton waves in the near-infrared spectral regime, *Optical Engineering*, 53:129801 (2014) (errata).

[60] M. E. Solano, G. D. Barber, A. Lakhtakia, M. Faryad, P. B. Monk, and T. E. Mallouk, Buffer layer between a planar optical concentrator and a solar cell, *AIP Advances*, 5:097150 (2015). 7

[61] L. Liu, G. D. Barber, M. V. Shuba, Y. Yuwen, A. Lakhtakia, T. E. Mallouk, and T. S. Mayer, Planar light concentration in micro-Si solar cells enabled by a metallic grating-photonic crystal architecture, *ACS Photonics*, 3:604–610 (2016). 7

[62] L. Liu, M. Faryad, A. S. Hall, G. D. Barber, S. Erten, T. E. Mallouk, A. Lakhtakia, and T. S. Mayer, Experimental excitation of multiple surface-plasmon-polariton waves and waveguide modes in a one-dimensional photonic crystal atop a two-dimensional metal grating, *Journal of Nanophotonics*, 9:093593 (2015). 7

[63] F.-J. Haug, K. Söderström, A. Naqavi, and C. Ballif, Excitation of guided-mode resonances in thin film silicon solar cells, *Materials Research Society Proceedings*, 1321:123–128 (2011). 7

[64] T. Khaleque and R. Magnusson, Light management through guided-mode resonances in thin-film silicon solar cells, *Journal of Nanophotonics*, 8:083995 (2014). 7

[65] T. H. Anderson, M. Faryad, T. G. Mackay, A. Lakhtakia, and R. Singh, Combined optical-electrical finite-element simulations of thin-film solar cells with homogeneous and nonhomogeneous intrinsic layers, *Journal of Photonics for Energy*, 6:025502 (2016). 7

[66] T. H. Anderson, T. G. Mackay, and A. Lakhtakia, Enhanced efficiency of Schottky-barrier solar cell with periodically nonhomogeneous indium gallium nitride layer, *Journal of Photonics for Energy*, 7:014502 (2017). 7

[67] J. A. Hutchby, High-efficiency graded band-gap $Al_xGa_{1-x}As$—GaAs solar cell, *Applied Physics Letters*, 26:457–459 (1975). 7

[68] I. M. Dharmadasa, Third generation multi-layer tandem solar cells for achieving high conversion efficiencies, *Solar Energy Materials and Solar Cells*, 85:293–300 (2005). 7

[69] I. M. Dharmadasa, A. A. Ojo, H. I. Salim, and R. Dharmadasa, Next generation solar cells based on graded bandgap device structures utilising rod-type nano-materials, *Energies*, 8:5440–5458 (2015). 7

[70] Y. Chen, P. Kivisaari, M.-E. Pistol, and N. Anttu, Optimization of the short-circuit current in an InP nanowire array solar cell through opto-electronic modeling, *Nanotechnology*, 27:435404 (2016). 7, 8

[71] R. Dewan, M. Marinkovic, R. Noriega, S. Phadke, S. A. Salleo, and D. Knipp, Light trapping in thin-film silicon solar cells with submicron surface texture, *Optics Express*, 17:23058–23065 (2009). 7

[72] V. E. Ferry, L. A. Sweatlock, D. Pacifici, and H. A. Atwater, Plasmonic nanostructure design for efficient light coupling into solar cells, *Nano Letters*, 8:4391–4397 (2008). 7

[73] H. A. Atwater and A. Polman, Plasmonics for improved photovoltaic devices, *Nature Materials*, 9:205–213 (2010). 7

[74] P. Baruch, A. D. Vos, P. T. Landsberg, and J. E. Parrott, On some thermodynamic aspects of photovoltaic solar energy conversion, *Solar Energy Materials and Solar Cells*, 36:201–222 (1995). 7

[75] J. Nelson, *The Physics of Solar Cells* (Imperial College Press, London, UK, 2003). 7, 8

[76] S. J. Fonash, *Solar Cell Device Physics*, 2nd ed. (Academic Press, Burlington, MA, 2010). 7, 8

[77] T. H. Anderson, B. J. Civiletti, P. B. Monk, and A. Lakhtakia, Coupled optoelectronic simulation and optimization of thin-film photovoltaic solar cells, *Journal of Computational Physics*, 407:109242 (2020). 8

T. H. Anderson, B. J. Civiletti, P. B. Monk, and A. Lakhtakia, Coupled optoelectronic simulation and optimization of thin-film photovoltaic solar cells, *Journal of Computational Physics*, 418:109561 (2020) (corrigendum).

[78] S. Adachi, *Earth-Abundant Materials for Solar Cells* (Wiley, Chichester, West Sussex, UK, 2015). 8

[79] R. Storn and K. Price, Differential evolution—a simple and efficient heuristic for global optimization over continuous spaces, *Journal of Global Optimization*, 11:341–359 (1997). 8

[80] S. Das and P. N. Suganthan, Differential evolution: A survey of the state-of-the-art, *IEEE Transactions on Evolutionary Computation*, 15:4–31 (2011). 8

[81] F. Ahmad, Optoelectronic modeling and optimization of graded-bandgap thin-film solar cells, Ph.D. Dissertation (The Pennsylvania State University, University Park, PA, 2020). 8

[82] J. A. Polo Jr., T. G. Mackay, and A. Lakhtakia, *Electromagnetic Surface Waves: A Modern Perspective* (Elsevier, Waltham, MA, 2013). 8

[83] T. G. Mackay and A. Lakhtakia, *The Transfer-Matrix Method in Electromagnetics and Optics* (Morgan & Claypool, San Rafael, CA, 2020). 8

[84] G. Fu, W. Qiu, and W. Zhang, An analysis of HDG methods for convection-dominated diffusion problems, *ESAIM: Mathematical Modelling and Numerical Analysis*, 49:225–256 (2015). 8

[85] C. Lehrenfeld, Hybrid discontinuous Galerkin methods for solving incompressible flow problems, Diplomingenieur Dissertation (Rheinisch-Westfaälischen Technischen Hochschule, Aachen, Germany, 2010). 8

[86] B. Cockburn, J. Gopalakrishnan, and R. Lazarov, Unified hybridization of discontinuous Galerkin, mixed, and continuous Galerkin methods for second order elliptic problems, *SIAM Journal on Numerical Analysis*, 47:1319–1365 (2009). 8

[87] D. Brinkman, K. Fellner, P. Markowich, and M.-T. Wolfram, A drift-diffusion-reaction model for excitonic photovoltaic bilayers: Asymptotic analysis and a 2-D HDG finite-element scheme, *Mathematical Models and Methods in Applied Sciences*, 23:839–872 (2013). 8

CHAPTER 2

Solar-Cell Optics

2.1 FREQUENCY-DOMAIN MAXWELL EQUATIONS

The electron-hole-pair generation rate G and the optical short-circuit current density J_{sc}^{opt} are calculated by knowing the electric field inside the solar cell in the optical regime. These calculations require the solution of the frequency-domain Maxwell equations [1, 2]

$$\nabla \times \mathbf{H}(\mathbf{r}, \omega) = -i\omega\mathbf{D}(\mathbf{r}, \omega), \qquad (2.1a)$$

$$\nabla \times \mathbf{E}(\mathbf{r}, \omega) = i\omega\mathbf{B}(\mathbf{r}, \omega), \qquad (2.1b)$$

$$\nabla \cdot \mathbf{D}(\mathbf{r}, \omega) = 0, \qquad (2.1c)$$

and

$$\nabla \cdot \mathbf{B}(\mathbf{r}, \omega) = 0, \qquad (2.1d)$$

where $i = \sqrt{-1}$, $\mathbf{r}$ is the position vector, ω is the angular frequency, $\mathbf{E}(\mathbf{r}, \omega)$ is the electric field phasor, $\mathbf{B}(\mathbf{r}, \omega)$ is the magnetic field phasor, $\mathbf{D}(\mathbf{r}, \omega)$ is the electric displacement field phasor, and $\mathbf{H}(\mathbf{r}, \omega)$ is the magnetic induction field phasor. For an isotropic dielectric medium, the constitutive relations are written as

$$\mathbf{D}(\mathbf{r}, \omega) = \varepsilon(\mathbf{r}, \omega)\mathbf{E}(\mathbf{r}) \qquad (2.2a)$$

and

$$\mathbf{B}(\mathbf{r}, \omega) = \mu_0\mathbf{H}(\mathbf{r}), \qquad (2.2b)$$

where $\varepsilon(\mathbf{r}, \omega)$ is the spatially varying permittivity and μ_0 is the permeability of free space.

Throughout this monograph, ε_0 is the permittivity of free space, $k_0 = \omega\sqrt{\varepsilon_0\mu_0}$ is the free-space wavenumber, $\lambda_0 = 2\pi/k_0$ is the free-space wavelength, and $\eta_0 = \sqrt{\mu_0/\varepsilon_0}$ is the intrinsic impedance of free space. The Cartesian coordinate system (x, y, z) is used, with unit vectors denoted by $\hat{\mathbf{u}}_x, \hat{\mathbf{u}}_y,$ and $\hat{\mathbf{u}}_z$. Vectors are denoted in boldface, whereas column vectors and matrixes are in boldface and decorated by an overhead cup.

2.2 RIGOROUS COUPLED-WAVE APPROACH

The rigorous coupled-wave approach (RCWA) is used to solve the frequency-domain Maxwell equations in a spatial domain with either one-dimensional (1D) or two-dimensional (2D) periodicity. It is a robust semi-analytical technique for the computation of plane-wave transmittances, reflectances, absorptances, and field profiles of diffraction gratings, resonant gratings,

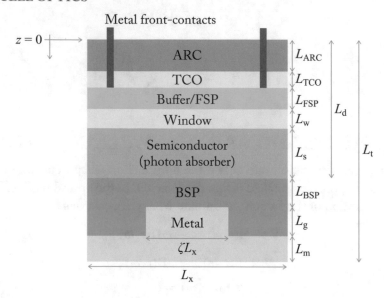

Figure 2.1: Schematic of the unit cell of a thin-film solar cell backed by a 1D periodically corrugated back reflector.

and photonic crystals [2–4]. This approach has the advantage of a shorter computational time of electromagnetic field phasors over other numerical techniques [5, 6] and is widely used for simulating solar cells with grating structures [7–10]. The theory developed in this section is valid for any thin-film solar cell backed by a 1D periodically corrugated back reflector, the theory for a 2D periodically corrugated reflector being available elsewhere [2]. The selection of 1D periodically corrugated back reflectors over 2D periodically corrugated back reflectors is due to lower computational time with the same benefits [11, 12].

The entire structure is translationally invariant along the y axis, so that only the xz plane needs to be considered, as shown schematically in Fig. 2.1. The solar cell occupies the region $\mathcal{X} : \{(x, z)| -\infty < x < \infty, 0 < z < L_t\}$, with the half spaces $z < 0$ and $z > L_t$ occupied by air. The reference unit cell is identified as $\mathcal{R} : \{(x, z)| -L_x/2 < x < L_x/2, 0 < z < L_t\}$, the back reflector being periodically corrugated along the x axis with period L_x.

The region $0 < z < L_{ARC}$ is occupied by either a single-layered or a multi-layered dielectric material which works as an anti-reflection coating (ARC) [13]. The region $L_{ARC} < z < L_{ARC} + L_{TCO}$ is occupied by a single layer of a transparent conducting oxide (TCO) to which the front electrode is connected. The region $L_{ARC} + L_{TCO} < z < L_{ARC} + L_{TCO} + L_{FSP}$ is occupied by a semiconductor material which acts as a buffer or front-surface passivation (FSP) layer [14]. The region $L_{ARC} + L_{TCO} + L_{FSP} < z < L_{ARC} + L_{TCO} + L_{FSP} + L_w$ is occupied by a semiconductor material which can be either single-layered or multi-layered to form a junction with the photon-absorbing layer of the solar cell. The region $L_{ARC} + L_{TCO} + L_{FSP} +$

$L_w < z < L_{ARC} + L_{TCO} + L_{FSP} + L_w + L_s = L_d$ is occupied by a semiconductor material which acts as a photon absorber and can be either single-layered or multi-layered. The photon-absorbing layer in the semiconductor region can either be homogeneous or graded along the thickness direction.

The region $L_d < z < L_d + L_{BSP}$ is a back-surface passivation (BSP) layer of thickness L_{BSP} used to reduce the back-surface recombination rate and avoid the deterioration of electrical properties of the photon-absorbing layer near the back electrode of the solar cell [14, 15].

The region $L_d + L_{BSP} < z < L_d + L_{BSP} + L_g$ is the grating region containing a metallic periodically corrugated back reflector. A rectangular periodically corrugated back reflector is shown as an example in Fig. 2.1. The periodically corrugated back reflector can be modeled through different functions according to the shape (e.g., rectangular, triangular, and sinusoidal) of the corrugations. The rectangular shape is considered for optimization throughout this book due to its simplicity. Also, this profile is justified as the corrugation shape does not significantly affect the maximization of efficiency of solar cells [16]. The metallic layer $L_t - L_m < z < L_t$ works both as a flat back reflector and the back electrode.

2.2.1   INCIDENT, REFLECTED, AND TRANSMITTED FIELDS

Suppose that an arbitrarily polarized plane wave is normally incident on the solar cell from the half-space $z < 0$. Its electric field phasor is denoted by

$$\mathbf{E}_{inc}(x, z, \lambda_0) = E_0 \frac{\bar{a}_p \, \hat{\mathbf{u}}_x + \bar{a}_s \, \hat{\mathbf{u}}_y}{\sqrt{\bar{a}_p^2 + \bar{a}_s^2}} \exp\left(i k_0 z\right) . \tag{2.3}$$

The complex-valued coefficients $\bar{a}_s$ and $\bar{a}_p$ determine the polarization state of the incident light; whereas, $\bar{a}_p = 1$ and $\bar{a}_s = 0$ for p-polarized light, $\bar{a}_s = 1$ and $\bar{a}_p = 0$ for s-polarized light [17, 18]. Although $\bar{a}_p = \bar{a}_s = 1$ indicates the linear polarization state, it is used for solar-cell calculations to represent the unpolarized state of direct sunlight. The amplitude E_0 of the incident electric field equals $4\sqrt{15\pi}$ V m^{-1} in order to correspond to unit incident power density (i.e., 1 W m^{-2}).

As a result of the metallic back reflector being periodic along the x axis, the electric and magnetic field phasors are represented everywhere as infinite series of Floquet harmonics [2, 18, 19]. Thus, the incident field phasors are

$$\left.\begin{array}{l} \mathbf{E}_{inc}(x, z, \lambda_0) = \displaystyle\sum_{m=-\infty}^{\infty} \left\{ \left[a_s^{(m)}\mathbf{s}^{(m)} + a_p^{(m)}\mathbf{p}_+^{(m)}\right] \exp\left[i\left(\kappa^{(m)}x + \alpha_0^{(m)}z\right)\right]\right\} \\[4mm] \mathbf{H}_{inc}(x, z, \lambda_0) = \eta_0^{-1}\displaystyle\sum_{m=-\infty}^{\infty} \left\{ \left[a_s^{(m)}\mathbf{p}_+^{(m)} - a_p^{(m)}\mathbf{s}^{(m)}\right] \exp\left[i\left(\kappa^{(m)}x + \alpha_0^{(m)}z\right)\right]\right\} \\[4mm] z < 0, \quad |x| < \infty, \end{array}\right\} \tag{2.4}$$

where the incidence amplitudes

$$a_s^{(m)} = \delta_{m0} \frac{\bar{a}_s E_0}{\sqrt{\bar{a}_p^2 + \bar{a}_s^2}} \tag{2.5a}$$

and

$$a_p^{(m)} = \delta_{m0} \frac{\bar{a}_p E_0}{\sqrt{\bar{a}_p^2 + \bar{a}_s^2}} \tag{2.5b}$$

involve the Kronecker delta

$$\delta_{mm'} = \begin{cases} 1, & m = m', \\ 0, & m \neq m'. \end{cases} \tag{2.6}$$

Other quantities involved in Eqs. (2.4) are

$$\left. \begin{aligned} \kappa^{(m)} &= m(2\pi/L_x) \\ s^{(m)} &= \hat{u}_y \\ p_+^{(m)}(\lambda_0) &= \frac{-\alpha_0^{(m)} \hat{u}_x + \kappa^{(m)} \hat{u}_z}{k_0} \\ \alpha_0^{(m)}(\lambda_0) &= +\sqrt{k_0^2 - (\kappa^{(m)})^2} \end{aligned} \right\} , \quad m \in \{0, \pm 1, \pm 2, \cdots\} . \tag{2.7}$$

The reflected field phasors are written as

$$\left. \begin{aligned} \mathbf{E}_{\text{ref}}(x, z, \lambda_0) &= \sum_{m=-\infty}^{\infty} \left\{ \left[r_s^{(m)} s^{(m)} + r_p^{(m)} p_-^{(m)} \right] \exp\left[i \left(\kappa^{(m)} x - \alpha_0^{(m)} z \right) \right] \right\} \\ \mathbf{H}_{\text{ref}}(x, z, \lambda_0) &= \eta_0^{-1} \sum_{m=-\infty}^{\infty} \left\{ \left[r_s^{(m)} p_-^{(m)} - r_p^{(m)} s^{(m)} \right] \exp\left[i \left(\kappa^{(m)} x - \alpha_0^{(m)} z \right) \right] \right\} \\ z < 0, \quad |x| < \infty, \end{aligned} \right\} , \tag{2.8}$$

where $r_s^{(m)}(\lambda_0)$ and $r_p^{(m)}(\lambda_0)$ are unknown reflection amplitudes and

$$p_-^{(m)}(\lambda_0) = \frac{\alpha_0^{(m)} \hat{u}_x + \kappa^{(m)} \hat{u}_z}{k_0} , \quad m \in \{0, \pm 1, \pm 2, \cdots\} . \tag{2.9}$$

Similarly, the transmitted field phasors are written as

$$\left. \begin{aligned} \mathbf{E}_{\text{tr}}(x, z, \lambda_0) &= \sum_{m=-\infty}^{\infty} \left(\left[t_s^{(m)} s^{(m)} + t_p^{(m)} p_+^{(m)} \right] \exp\left\{ i \left[\kappa^{(m)} x + \alpha_0^{(m)} (z - L_t) \right] \right\} \right) \\ \mathbf{H}_{\text{tr}}(x, z, \lambda_0) &= \eta_0^{-1} \sum_{m=-\infty}^{\infty} \left(\left[t_s^{(m)} p_+^{(m)} - t_p^{(m)} s^{(m)} \right] \exp\left\{ i \left[\kappa^{(m)} x + \alpha_0^{(m)} (z - L_t) \right] \right\} \right) \\ z > L_t, \quad |x| < \infty, \end{aligned} \right\} , \tag{2.10}$$

where $t_s^{(m)}(\lambda_0)$ and $t_p^{(m)}(\lambda_0)$ are unknown transmission amplitudes.

2.2.2 FIELDS INSIDE THE SOLAR CELL

The variations of the field phasors in the solar cell along the x axis are best expressed as the Fourier series [2, 18]

$$
\left.
\begin{aligned}
\mathbf{E}(x, z, \lambda_0) &= \sum_{m=-\infty}^{\infty} \left[\mathbf{e}^{(m)}(z, \lambda_0) \exp\left(i\kappa^{(m)} x \right) \right] \\
\mathbf{H}(x, z, \lambda_0) &= \sum_{m=-\infty}^{\infty} \left[\mathbf{h}^{(m)}(z, \lambda_0) \exp\left(i\kappa^{(m)} x \right) \right]
\end{aligned}
\right\}, \quad 0 < z < L_{\mathrm{t}}, \quad |x| < \infty, \quad (2.11)
$$

where

$$
\mathbf{e}^{(m)}(z, \lambda_0) = e_{\mathrm{x}}^{(m)}(z, \lambda_0)\hat{\mathbf{u}}_x + e_{\mathrm{y}}^{(m)}(z, \lambda_0)\hat{\mathbf{u}}_y + e_{\mathrm{z}}^{(m)}(z, \lambda_0)\hat{\mathbf{u}}_z \quad (2.12a)
$$

and

$$
\mathbf{h}^{(m)}(z, \lambda_0) = h_{\mathrm{x}}^{(m)}(z, \lambda_0)\hat{\mathbf{u}}_x + h_{\mathrm{y}}^{(m)}(z, \lambda_0)\hat{\mathbf{u}}_y + h_{\mathrm{z}}^{(m)}(z, \lambda_0)\hat{\mathbf{u}}_z \quad (2.12b)
$$

are Fourier coefficients. Finally, the permittivity $\varepsilon(x, z, \lambda_0)$ inside the solar cell is represented by the Fourier series

$$
\varepsilon(x, z, \lambda_0) = \sum_{m=-\infty}^{\infty} \left[\varepsilon^{(m)}(z, \lambda_0) \exp\left(i\kappa^{(m)} x \right) \right], \quad 0 < z < L_{\mathrm{t}}, \quad |x| < \infty, \quad (2.13)
$$

where $\varepsilon^{(m)}(z, \lambda_0)$ are Fourier coefficients. Substitution of Eqs. (2.2a), (2.11), and (2.13) into Eqs. (2.1a) and (2.1b) results in the four ordinary differential equations

$$
\frac{d}{dz} e_{\mathrm{x}}^{(m)}(z, \lambda_0) - i\kappa^{(m)} e_{\mathrm{z}}^{(m)}(z, \lambda_0) = i\omega\mu_0 h_{\mathrm{y}}^{(m)}(z, \lambda_0), \quad (2.14a)
$$

$$
\frac{d}{dz} e_{\mathrm{y}}^{(m)}(z, \lambda_0) = -i\omega\mu_0 h_{\mathrm{x}}^{(m)}(z, \lambda_0), \quad (2.14b)
$$

$$
\frac{d}{dz} h_{\mathrm{x}}^{(m)}(z, \lambda_0) - i\kappa^{(m)} h_{\mathrm{z}}^{(m)}(z, \lambda_0) = -i\omega \sum_{\ell=-\infty}^{\infty} \varepsilon^{(m-\ell)}(z, \lambda_0) e_{\mathrm{y}}^{(\ell)}(z, \lambda_0), \quad (2.14c)
$$

and

$$
\frac{d}{dz} h_{\mathrm{y}}^{(m)}(z, \lambda_0) = i\omega \sum_{\ell=-\infty}^{\infty} \varepsilon^{(m-\ell)}(z, \lambda_0) e_{\mathrm{x}}^{(\ell)}(z, \lambda_0), \quad (2.14d)
$$

as well as the two algebraic equations

$$
\kappa^{(m)} e_{\mathrm{y}}^{(m)}(z, \lambda_0) = \omega\mu_0 h_{\mathrm{z}}^{(m)}(z, \lambda_0) \quad (2.14e)
$$

and

$$
\kappa^{(m)} h_{\mathrm{y}}^{(m)}(z, \lambda_0) = -\omega \sum_{\ell=-\infty}^{\infty} \varepsilon^{(m-\ell)}(z, \lambda_0) e_{\mathrm{z}}^{(\ell)}(z, \lambda_0). \quad (2.14f)
$$

All six equations hold true inside the solar cell (i.e., for $0 < z < L_{\mathrm{t}}$ and $|x| < \infty$).

2.2.3 TRUNCATED EXPRESSIONS

Computational tractability requires the expansions in Eqs. (2.4), (2.8), (2.10), (2.11), (2.13), and (2.14) to be truncated to include only $m \in \{-M_t, ..., 0, ..., M_t\}$ with $M_t \geq 0$. The phasor amplitudes of the s- and p-polarized components of the incident, reflected, and transmitted fields are written as $2(2M_t + 1)$-column vectors as

$$\check{\mathbf{A}} = \left[a_s^{(-M_t)}, a_s^{(-M_t+1)}, \cdots, a_s^{(M_t-1)}, a_s^{(M_t)}, a_p^{(-M_t)}, a_p^{(-M_t+1)}, \cdots, a_p^{(M_t-1)}, a_p^{(M_t)} \right]^T , \quad (2.15a)$$

$$\check{\mathbf{R}}(\lambda_0) = \left[r_s^{(-M_t)}, r_s^{(-M_t+1)}, \cdots, r_s^{(M_t-1)}, r_s^{(M_t)}, r_p^{(-M_t)}, r_p^{(-M_t+1)}, \cdots, r_p^{(M_t-1)}, r_p^{(M_t)} \right]^T , \quad (2.15b)$$

and

$$\check{\mathbf{T}}(\lambda_0) = \left[t_s^{(-M_t)}, t_s^{(-M_t+1)}, \cdots, t_s^{(M_t-1)}, t_s^{(M_t)}, t_p^{(-M_t)}, t_p^{(-M_t+1)}, \cdots, t_p^{(M_t-1)}, t_p^{(M_t)} \right]^T , \quad (2.15c)$$

where the superscript T denotes the transpose. In addition, the six $(2M_t + 1)$-column vectors

$$\check{\mathbf{e}}_\sigma(z, \lambda_0) = \left[e_\sigma^{(-M_t)}, e_\sigma^{(-M_t+1)}, ..., e_\sigma^{(M_t-1)}, e_\sigma^{(M_t)} \right]^T , \quad \sigma \in \{x, y, z\}, \quad (2.16a)$$

and

$$\check{\mathbf{h}}_\sigma(z, \lambda_0) = \left[h_\sigma^{(-M_t)}, h_\sigma^{(-M_t+1)}, ..., h_\sigma^{(M_t-1)}, h_\sigma^{(M_t)} \right]^T , \quad \sigma \in \{x, y, z\}, \quad (2.16b)$$

are set up to represent the Fourier coefficients of $\mathbf{E}$ and $\mathbf{H}$ inside the solar cell. The Toeplitz matrix [20]

$$\check{\boldsymbol{\varepsilon}}(z, \lambda_0) = \begin{bmatrix} \varepsilon^{(0)} & \varepsilon^{(-1)} & \cdots & \varepsilon^{(-2M_t+1)} & \varepsilon^{(-2M_t)} \\ \varepsilon^{(1)} & \varepsilon^{(0)} & \cdots & \varepsilon^{(-2M_t+2)} & \varepsilon^{(-2M_t+1)} \\ \cdots & \cdots & \cdots & \cdots & \cdots \\ \varepsilon^{(2M_t-1)} & \varepsilon^{(2M_t-2)} & \cdots & \varepsilon^{(0)} & \varepsilon^{(-1)} \\ \varepsilon^{(2M_t)} & \varepsilon^{(2M_t-1)} & \cdots & \varepsilon^{(1)} & \varepsilon^{(0)} \end{bmatrix} \quad (2.17)$$

contains the Fourier coefficients appearing in Eq. (2.13). Finally, the $(2M_t + 1) \times (2M_t + 1)$ Fourier-wavenumber matrix is set up as

$$\check{\mathbf{K}} = \text{diag} \left[\kappa^{(-M_t)}, \kappa^{(-M_t+1)}, \cdots, \kappa^{(M_t-1)}, \kappa^{(M_t)} \right] . \quad (2.18)$$

2.2.4 MATRIX ORDINARY DIFFERENTIAL EQUATION

Equations (2.14e) and (2.14f) give rise to the matrix algebraic equations

$$\check{\mathbf{h}}_z(z, \lambda_0) = (\omega\mu_0)^{-1}\check{\mathbf{K}} \cdot \check{\mathbf{e}}_y(z, \lambda_0) \quad (2.19a)$$

and

$$\breve{e}_z(z, \lambda_0) = - [\omega \breve{\boldsymbol{\varepsilon}}(z, \lambda_0)]^{-1} \cdot \breve{\mathbf{K}} \cdot \breve{\mathbf{h}}_y(z, \lambda_0), \qquad (2.19b)$$

respectively. Equations (2.1a) and (2.1b) thereafter yield the matrix ordinary differential equation [2, 18]

$$\frac{d}{dz}\breve{\mathbf{f}}(z, \lambda_0) = i \breve{\mathbf{P}}(z, \lambda_0) \cdot \breve{\mathbf{f}}(z, \lambda_0), \qquad 0 < z < L_t, \qquad (2.20)$$

satisfied by the $4(2M_t + 1)$-column vector

$$\breve{\mathbf{f}}(z, \lambda_0) = \begin{bmatrix} \breve{e}_x(z, \lambda_0) \\ \breve{e}_y(z, \lambda_0) \\ \breve{h}_x(z, \lambda_0) \\ \breve{h}_y(z, \lambda_0) \end{bmatrix}. \qquad (2.21)$$

The $4(2M_t + 1) \times 4(2M_t + 1)$ matrix

$$\breve{\mathbf{P}}(z, \lambda_0) = \omega \begin{bmatrix} \breve{\mathbf{O}} & \breve{\mathbf{O}} & \breve{\mathbf{O}} & \mu_0 \breve{\mathbf{I}} \\ \breve{\mathbf{O}} & \breve{\mathbf{O}} & -\mu_0 \breve{\mathbf{I}} & \breve{\mathbf{O}} \\ \breve{\mathbf{O}} & -\breve{\boldsymbol{\varepsilon}}(z, \lambda_0) & \breve{\mathbf{O}} & \breve{\mathbf{O}} \\ \breve{\boldsymbol{\varepsilon}}(z, \lambda_0) & \breve{\mathbf{O}} & \breve{\mathbf{O}} & \breve{\mathbf{O}} \end{bmatrix}$$

$$+ \frac{1}{\omega} \begin{bmatrix} \breve{\mathbf{O}} & \breve{\mathbf{O}} & \breve{\mathbf{O}} & -\breve{\mathbf{K}} \cdot [\breve{\boldsymbol{\varepsilon}}(z, \lambda_0)]^{-1} \cdot \breve{\mathbf{K}} \\ \breve{\mathbf{O}} & \breve{\mathbf{O}} & \breve{\mathbf{O}} & \breve{\mathbf{O}} \\ \breve{\mathbf{O}} & \mu_0^{-1} \breve{\mathbf{K}} \cdot \breve{\mathbf{K}} & \breve{\mathbf{O}} & \breve{\mathbf{O}} \\ \breve{\mathbf{O}} & \breve{\mathbf{O}} & \breve{\mathbf{O}} & \breve{\mathbf{O}} \end{bmatrix} \qquad (2.22)$$

in Eq. (2.20) contains $\breve{\mathbf{O}}$ as the $(2M_t + 1) \times (2M_t + 1)$ null matrix and $\breve{\mathbf{I}}$ as the $(2M_t + 1) \times (2M_t + 1)$ identity matrix.

Equation (2.20) has to be solved to conform to the the boundary values

$$\breve{\mathbf{f}}(0^-, \lambda_0) = \begin{bmatrix} \breve{\mathbf{O}} & -k_0^{-1} \breve{\boldsymbol{\alpha}}_0(\lambda_0) & \breve{\mathbf{O}} & k_0^{-1} \breve{\boldsymbol{\alpha}}_0(\lambda_0) \\ \breve{\mathbf{I}} & \breve{\mathbf{O}} & \breve{\mathbf{I}} & \breve{\mathbf{O}} \\ -\eta_0^{-1} k_0^{-1} \breve{\boldsymbol{\alpha}}_0(\lambda_0) & \breve{\mathbf{O}} & \eta_0^{-1} k_0^{-1} \breve{\boldsymbol{\alpha}}_0(\lambda_0) & \breve{\mathbf{O}} \\ \breve{\mathbf{O}} & -\eta_0^{-1} \breve{\mathbf{I}} & \breve{\mathbf{O}} & -\eta_0^{-1} \breve{\mathbf{I}} \end{bmatrix} \cdot \begin{bmatrix} \breve{\mathbf{A}} \\ \breve{\mathbf{R}}(\lambda_0) \end{bmatrix}$$

$$(2.23a)$$

and

$$\breve{\mathbf{f}}(L_t^+, \lambda_0) = \begin{bmatrix} \breve{\mathbf{O}} & -k_0^{-1}\,\breve{\boldsymbol{\alpha}}_0(\lambda_0) \\ \breve{\mathbf{I}} & \breve{\mathbf{O}} \\ -\eta_0^{-1}\,k_0^{-1}\,\breve{\boldsymbol{\alpha}}_0(\lambda_0) & \breve{\mathbf{O}} \\ \breve{\mathbf{O}} & -\eta_0^{-1}\,\breve{\mathbf{I}} \end{bmatrix} \cdot \begin{bmatrix} \breve{\mathbf{T}}(\lambda_0) \end{bmatrix}, \qquad (2.23b)$$

with the $(2M_t + 1) \times (2M_t + 1)$ diagonal matrix

$$\breve{\boldsymbol{\alpha}}_0(\lambda_0) = \mathrm{diag}\left[\alpha_0^{(-M_t)}, \alpha_0^{(-M_t+1)}, \cdots, \alpha_0^{(M_t-1)}, \alpha_0^{(M_t)}\right]. \qquad (2.24)$$

Here and hereafter, $\breve{\mathbf{f}}(z^{\pm}, \lambda_0)$ is the limiting value of $\breve{\mathbf{f}}(z \pm \delta_z, \lambda_0)$ as the non-negative δ_z goes to 0. For convenience, the boundary values are recast compactly as

$$\breve{\mathbf{f}}(0^-, \lambda_0) = \begin{bmatrix} \breve{\mathbf{Y}}_e^{\mathrm{inc}}(\lambda_0) & \breve{\mathbf{Y}}_e^{\mathrm{ref}}(\lambda_0) \\ \breve{\mathbf{Y}}_h^{\mathrm{inc}}(\lambda_0) & \breve{\mathbf{Y}}_h^{\mathrm{ref}}(\lambda_0) \end{bmatrix} \cdot \begin{bmatrix} \breve{\mathbf{A}} \\ \breve{\mathbf{R}}(\lambda_0) \end{bmatrix} \qquad (2.25a)$$

and

$$\breve{\mathbf{f}}(L_t^+, \lambda_0) = \begin{bmatrix} \breve{\mathbf{Y}}_e^{\mathrm{tr}}(\lambda_0) \\ \breve{\mathbf{Y}}_h^{\mathrm{tr}}(\lambda_0) \end{bmatrix} \cdot \breve{\mathbf{T}}(\lambda_0). \qquad (2.25b)$$

The $2(2M_t+1) \times 2(2M_t+1)$ matrixes $\breve{\mathbf{Y}}_e^{\mathrm{inc}}(\lambda_0)$, etc., can be synthesized from Eqs. (2.23a) and (2.23b) by inspection.

2.2.5   SOLUTION OF MATRIX ORDINARY DIFFERENTIAL EQUATION

To solve Eq. (2.20), the region $\mathcal{R}$ is partitioned into a sufficiently large number N_s of thin slices along the z axis [2, 5, 6]. The ℓth slice is bounded by the planes $z = z_{\ell-1}$ and $z = z_\ell$, $\ell \in [1, N_s]$, where $z_0 = 0$ and $z_{N_s} = L_t$. In the ℓth slice, the matrix $\breve{\mathbf{P}}(z, \lambda_0)$ is approximated by the uniform matrix

$$\breve{\mathbf{P}}^{(\ell)}(\lambda_0) = \breve{\mathbf{P}}\left(\frac{z_{\ell-1} + z_\ell}{2}, \lambda_0\right), \qquad \ell \in [1, N_s], \qquad (2.26)$$

and the matrix

$$\breve{\mathbf{W}}^{(\ell)}(\lambda_0) = \exp\left[i(z_\ell - z_{\ell-1})\,\breve{\mathbf{P}}^{(\ell)}(\lambda_0)\right], \qquad \ell \in [1, N_s], \qquad (2.27)$$

is formulated. Then,

$$\breve{\mathbf{f}}(L_t^-, \lambda_0) \approx \breve{\mathbf{W}}^{(N_s)}(\lambda_0) \cdot \breve{\mathbf{W}}^{(N_s-1)}(\lambda_0) \cdot \ldots \cdot \breve{\mathbf{W}}^{(2)}(\lambda_0) \cdot \breve{\mathbf{W}}^{(1)}(\lambda_0) \cdot \breve{\mathbf{f}}(0^+, \lambda_0), \qquad (2.28)$$

provided that every slice is sufficiently thin.

The standard boundary conditions being

$$\breve{\mathbf{f}}(0^-, \lambda_0) = \breve{\mathbf{f}}(0^+, \lambda_0) \tag{2.29a}$$

and

$$\breve{\mathbf{f}}(L_t^-, \lambda_0) = \breve{\mathbf{f}}(L_t^+, \lambda_0)\,, \tag{2.29b}$$

Eqs. (2.25a), (2.25b), and (2.28) lead to the algebraic equation

$$\begin{bmatrix} \breve{\mathbf{Y}}_e^{tr} \\ \breve{\mathbf{Y}}_h^{tr} \end{bmatrix} \cdot \breve{\mathbf{T}} = \breve{\mathbf{W}}^{(N_s)} \cdot \breve{\mathbf{W}}^{(N_s-1)} \cdot \ldots \cdot \breve{\mathbf{W}}^{(2)} \cdot \breve{\mathbf{W}}^{(1)} \cdot \begin{bmatrix} \breve{\mathbf{Y}}_e^{inc} & \breve{\mathbf{Y}}_e^{ref} \\ \breve{\mathbf{Y}}_h^{inc} & \breve{\mathbf{Y}}_h^{ref} \end{bmatrix} \cdot \begin{bmatrix} \breve{\mathbf{A}} \\ \breve{\mathbf{R}} \end{bmatrix}, \tag{2.30}$$

which may be solved for $\breve{\mathbf{R}}(\lambda_0)$ and $\breve{\mathbf{T}}(\lambda_0)$ using standard matrix techniques [21].

2.2.6   STABLE ALGORITHM

However, the application of a standard matrix-inversion technique such as the Gauss elimination technique [21] to solve Eq. (2.20) is prone to numerical problems. A stable algorithm based on matrix diagonalization is available to avoid the matrix-inversion problems in the RCWA [2, 18].[1]

The stable algorithm requires $\breve{\mathbf{P}}^{(\ell)}$ to be diagonalizable for every $\ell \in [1, N_s]$, i.e.,

$$\breve{\mathbf{P}}^{(\ell)} = \breve{\mathbf{V}}^{(\ell)} \cdot \breve{\mathbf{G}}^{(\ell)} \cdot [\breve{\mathbf{V}}^{(\ell)}]^{-1}\,, \qquad \ell \in [1, N_s]\,, \tag{2.31}$$

where the diagonal matrix $\breve{\mathbf{G}}^{(\ell)}$ contains the eigenvalues of $\breve{\mathbf{P}}^{(\ell)}$ in decreasing order of the imaginary part, and $\breve{\mathbf{V}}^{(\ell)}$ is a square matrix comprising the eigenvectors of $\breve{\mathbf{P}}^{(\ell)}$ as its columns, arranged so that each eigenvector is in the same position as the corresponding eigenvalue on the diagonal of $\breve{\mathbf{G}}^{(\ell)}$. Accordingly, Eq. (2.20) yields

$$\breve{\mathbf{f}}(z_{\ell-1}) = \breve{\mathbf{V}}^{(\ell)} \cdot \exp\left[-i\,(z_\ell - z_{\ell-1})\,\breve{\mathbf{G}}^{(\ell)}\right] \cdot [\breve{\mathbf{V}}^{(\ell)}]^{-1} \cdot \breve{\mathbf{f}}(z_\ell)\,, \qquad \ell \in [1, N_s]\,. \tag{2.32}$$

Given that the solar cell is made of isotropic dielectric materials exclusively, the diagonalizability requirement is not onerous [22, 23].

A set of auxiliary $2(2M_t + 1)$- column vectors $\breve{\mathbf{T}}^{(\ell)}$ and auxiliary transmission matrixes $\breve{\mathbf{Z}}^{(\ell)}$ of size $4(2M_t+1) \times 2(2M_t+1)$ are postulated to satisfy the relations

$$\breve{\mathbf{f}}(z_\ell) = \breve{\mathbf{Z}}^{(\ell)} \cdot \breve{\mathbf{T}}^{(\ell)}\,, \qquad \ell \in [1, N_s]\,, \tag{2.33}$$

where the matrix

$$\breve{\mathbf{Z}}^{(N_s)} = \begin{bmatrix} \breve{\mathbf{Y}}_e^{tr} \\ \breve{\mathbf{Y}}_h^{tr} \end{bmatrix} \tag{2.34a}$$

---

[1]Dependences on λ_0 are left unstated in Section 2.2.6 to avoid notational clutter.

and the column vector

$$\check{\mathbf{T}}^{(N_s)} = \check{\mathbf{T}}.$$ (2.34b)

Substitution of Eq. (2.33) into Eq. (2.32) yields

$$\check{\mathbf{Z}}^{(\ell-1)} \cdot \check{\mathbf{T}}^{(\ell-1)} = \check{\mathbf{V}}^{(\ell)} \cdot$$

$$\left[\begin{array}{cc} \exp\left[-i\,(z_\ell - z_{\ell-1})\,\check{\mathbf{G}}^{(\ell)}_{\text{upper}}\right] & \check{\mathbf{0}} \\ \check{\mathbf{0}} & \exp\left[-i\,(z_\ell - z_{\ell-1})\,\check{\mathbf{G}}^{(\ell)}_{\text{lower}}\right] \end{array} \right]$$

$$\cdot \left[\check{\mathbf{V}}^{(\ell)}\right]^{-1} \cdot \check{\mathbf{T}}^{(\ell)} \cdot \check{\mathbf{Z}}^{(\ell)},$$ (2.35)

where $\check{\mathbf{0}}$ is the $2(2M_t + 1) \times 2(2M_t + 1)$ null matrix. The $2(2M_t + 1) \times 2(2M_t + 1)$ matrixes $\check{\mathbf{G}}^{(\ell)}_{\text{upper}}$ and $\check{\mathbf{G}}^{(\ell)}_{\text{lower}}$ are the upper and lower diagonal submatrixes of the $4(2M_t + 1) \times 4(2M_t + 1)$ matrix $\check{\mathbf{G}}^{(\ell)}$, respectively.

Next, the $2(2M_t + 1) \times 2(2M_t + 1)$ matrixes $\check{\mathbf{X}}^{(\ell)}_{\text{upper}}$ and $\check{\mathbf{X}}^{(\ell)}_{\text{lower}}$ are defined via

$$\left[\begin{array}{c} \check{\mathbf{X}}^{(\ell)}_{\text{upper}} \\ \check{\mathbf{X}}^{(\ell)}_{\text{lower}} \end{array} \right] = \left[\check{\mathbf{V}}^{(\ell)}\right]^{-1} \cdot \check{\mathbf{Z}}^{(\ell)}$$ (2.36a)

and the recurrence relation

$$\check{\mathbf{T}}^{(\ell-1)} = \exp\left[-i\,(z_\ell - z_{\ell-1})\,\check{\mathbf{G}}^{(\ell)}_{\text{upper}}\right] \cdot \check{\mathbf{X}}^{(\ell)}_{\text{upper}} \cdot \check{\mathbf{T}}^{(\ell)}$$ (2.36b)

is postulated. Now, substitution of Eq. (2.36b) in Eq. (2.35) leads to the relation

$$\check{\mathbf{Z}}^{(\ell-1)} = \check{\mathbf{V}}^{(\ell)} \cdot \left[\begin{array}{c} \check{\mathbf{1}} \\ \check{\mathbf{U}}^{(\ell)} \end{array} \right], \qquad \ell \in [1, N_s],$$ (2.37)

where $\check{\mathbf{1}}$ is the $2(2M_t + 1) \times 2(2M_t + 1)$ identity matrix and

$$\check{\mathbf{U}}^{(\ell)} = \exp\left[-i\,(z_\ell - z_{\ell-1})\,\check{\mathbf{G}}^{(\ell)}_{\text{lower}}\right] \cdot \check{\mathbf{X}}^{(\ell)}_{\text{lower}} \cdot \left(\check{\mathbf{X}}^{(\ell)}_{\text{upper}}\right)^{-1}$$

$$\cdot \exp\left[i\,(z_\ell - z_{\ell-1})\,\check{\mathbf{G}}^{(\ell)}_{\text{upper}}\right], \qquad \ell \in [1, N_s].$$ (2.38)

After using Eqs. (2.36a) and (2.37) repeatedly, $\check{\mathbf{Z}}^{(\ell)}$ is found in terms of $\check{\mathbf{Z}}^{(N_s)}$ $\forall \ell \in [1, N_s - 1]$. Then, $\check{\mathbf{Z}}^{(0)}$ is further partitioned as

$$\check{\mathbf{Z}}^{(0)} = \left[\begin{array}{c} \check{\mathbf{Z}}^{(0)}_{\text{upper}} \\ \check{\mathbf{Z}}^{(0)}_{\text{lower}} \end{array} \right].$$ (2.39)

Setting the right side of Eq. (2.33) with $\ell = 0$ equal to the right side of Eq. (2.25a) yields $\check{\mathbf{R}}$ and $\check{\mathbf{T}}^{(0)}$ as follows:

$$
\left[\begin{array}{c} \check{\mathbf{T}}^{(0)} \\ \check{\mathbf{R}} \end{array} \right] = \left[\begin{array}{cc} \check{\mathbf{Z}}^{(0)}_{\text{upper}} & -\check{\mathbf{Y}}^{\text{ref}}_e \\ \check{\mathbf{Z}}^{(0)}_{\text{lower}} & -\check{\mathbf{Y}}^{\text{ref}}_h \end{array} \right]^{-1} \cdot \left[\begin{array}{c} \check{\mathbf{Y}}^{\text{inc}}_e \\ \check{\mathbf{Y}}^{\text{inc}}_h \end{array} \right] \cdot \check{\mathbf{A}} . \tag{2.40}
$$

After $\check{\mathbf{T}}^{(0)}$ is known, $\check{\mathbf{T}}^{(\ell)}$ is found for every $\ell \in [1, N_s]$ by reversing the sense of iterations in Eq. (2.36b). Thus, the stable RCWA algorithm yields both $\check{\mathbf{T}} = \check{\mathbf{T}}^{(N_s)}$ and $\check{\mathbf{R}}$. A Mathematica® program is available elsewhere [2].

Finally, the z-directed components of the magnetic and electric field phasors in the device can be obtained through Eqs. (2.19a) and (2.19b), respectively. Thus, the electric field phasor $\check{\mathbf{E}}(x, y, \lambda_0)$ can be determined everywhere inside the solar cell.

2.2.7   REFLECTANCES, TRANSMITTANCES, AND ABSORPTANCES

As the plane of propagation of the incident plane wave is the same as the grating plane and all the materials are isotropic, no depolarization can occur on reflection and transmission [2]. Hence, the reflection coefficients

$$
\left. \begin{array}{l} r^{(m)}_{\text{ss}}(\lambda_0) = r^{(m)}_{\text{s}}(\lambda_0)/a^{(0)}_{\text{s}} \\ r^{(m)}_{\text{pp}}(\lambda_0) = r^{(m)}_{\text{p}}(\lambda_0)/a^{(0)}_{\text{p}} \end{array} \right\} , \quad m \in [-M_t, M_t] , \tag{2.41}
$$

and the transmission coefficients

$$
\left. \begin{array}{l} t^{(m)}_{\text{ss}}(\lambda_0) = t^{(m)}_{\text{s}}(\lambda_0)/a^{(0)}_{\text{s}} \\ t^{(m)}_{\text{pp}}(\lambda_0) = t^{(m)}_{\text{p}}(\lambda_0)/a^{(0)}_{\text{p}} \end{array} \right\} , \quad m \in [-M_t, M_t] , \tag{2.42}
$$

are defined to optically characterize the solar cell. Coefficients of order $m = 0$ are classified as specular, whereas coefficients of all other orders are classified as nonspecular.

The reflectances of order m are defined as

$$
\left. \begin{array}{l} R^{(m)}_{\text{ss}}(\lambda_0) = \text{Re}\left[\alpha^{(m)}_0(\lambda_0)/k_0\right] \left|r^{(m)}_{\text{ss}}(\lambda_0)\right|^2 \in [0, 1] \\ R^{(m)}_{\text{pp}}(\lambda_0) = \text{Re}\left[\alpha^{(m)}_0(\lambda_0)/k_0\right] \left|r^{(m)}_{\text{pp}}(\lambda_0)\right|^2 \in [0, 1] \end{array} \right\} , \quad m \in [-M_t, M_t] , \tag{2.43}
$$

and the transmittances of order m as

$$
\left. \begin{array}{l} T^{(m)}_{\text{ss}}(\lambda_0) = \text{Re}\left[\alpha^{(m)}_0(\lambda_0)/k_0\right] \left|t^{(m)}_{\text{ss}}(\lambda_0)\right|^2 \in [0, 1] \\ T^{(m)}_{\text{pp}}(\lambda_0) = \text{Re}\left[\alpha^{(m)}_0(\lambda_0)/k_0\right] \left|t^{(m)}_{\text{pp}}(\lambda_0)\right|^2 \in [0, 1] \end{array} \right\} , \quad m \in [-M_t, M_t] . \tag{2.44}
$$

The absorptances for s- and p-polarized incidence conditions, respectively, can be determined as

$$
\left.\begin{aligned}
A_\mathrm{s}(\lambda_0) &= 1 - \sum_{m=-M_t}^{M_t} \left[R_\mathrm{ss}^{(m)}(\lambda_0) + T_\mathrm{ss}^{(m)}(\lambda_0) \right] \in [0, 1] \\
A_\mathrm{p}(\lambda_0) &= 1 - \sum_{m=-M_t}^{M_t} \left[R_\mathrm{pp}^{(m)}(\lambda_0) + T_\mathrm{pp}^{(m)}(\lambda_0) \right] \in [0, 1]
\end{aligned}\right\} ,
\tag{2.45}
$$

by virtue of the principle of conservation of energy. Maximization of the polarization-averaged absorptance

$$
\langle A(\lambda_0) \rangle_\mathrm{pol} = \frac{1}{2} \left[A_\mathrm{s}(\lambda_0) + A_\mathrm{p}(\lambda_0) \right]
\tag{2.46}
$$

is not useful for designing solar cells because it includes absorption in the metallic back reflector as well as in other non-semiconducting parts of the solar cell.

When the metal thickness L_m is chosen large enough for maximal reflection, the RCWA must yield $T_\mathrm{ss}^{(m)} \simeq 0$ and $T_\mathrm{pp}^{(m)} \simeq 0$ for all $m \in [-M_t, M_t]$. The adequate value of the truncation parameter M_t has to be determined by numerical computation. As M_t is increased in increments of unity, the convergence of the reflectances $R_\mathrm{ss}^{(m)}$ and $R_\mathrm{pp}^{(m)}$ must be checked until satisfactory.

Mathematical analyses of the convergence of the RCWA have been made for both s- and p-polarized light [5, 6]. Convergence rates have been proved as M_t and N_s simultaneously increase under two assumptions: that the permittivity is real and that it is either an increasing or a decreasing function of z in the non-metallic region of the structure. Convergence will still occur without these assumptions but could be slowed by resonances.

2.3 ELECTRON-HOLE-PAIR GENERATION

At any location inside the solar cell, the absorption rate of monochromatic energy per unit volume is given by

$$
Q(x, z, \lambda_0) = \frac{1}{2} \omega \, \mathrm{Im}\{\varepsilon(x, z, \lambda_0)\} \, |\mathbf{E}(x, z, \lambda_0)|^2 .
\tag{2.47}
$$

With the assumption that every absorbed photon with energy greater than the bandgap excites an electron-hole pair, the electron-hole-pair generation rate within a semiconductor layer of the solar cell is

$$
G(x, z) = \frac{2\eta_0}{\hbar E_0^2} \int_{\lambda_{0,\mathrm{min}}}^{\lambda_{0,\mathrm{max}}} \omega^{-1} Q(x, z, \lambda_0) S(\lambda_0) \, d\lambda_0 ,
\tag{2.48}
$$

where $S(\lambda_0)$ is the solar irradiance, $\hbar = 1.054 \times 10^{-34}$ J s is the reduced Planck constant, $\lambda_{0,\mathrm{min}} = 300$ nm, and $\lambda_{0,\mathrm{max}} = (1240/\mathrm{E_a})$ nm with $\mathrm{E_a}$ specified in eV being the minimum bandgap of the semiconductor. The optical short-circuit current density is then given by

$$
J_\mathrm{sc}^\mathrm{opt} = \frac{q_\mathrm{e}}{L_\mathrm{x}} \iint_{\mathcal{R}_\mathrm{sc}} G(x, z) \, dx \, dz ,
\tag{2.49}
$$

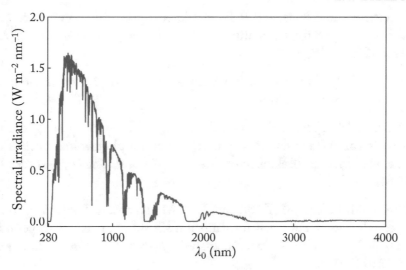

Figure 2.2: Irradiance $S(\lambda_0)$ as a function of free-space wavelength λ_0 in the AM1.5G solar spectrum [24].

where $q_e = 1.602 \times 10^{-19}$ C and $\mathcal{R}_{sc}$ is the portion of $\mathcal{R}$ occupied by semiconductors.

Solar irradiation is accounted for solar-cell calculations through the solar irradiance function $S(\lambda_0)$. It contains the contribution of direct as well as diffuse sunlight. For terrestrial solar cells, the AM1.5G solar spectrum [24], shown graphically in Fig. 2.2, is used as a standard.

2.4   BIBLIOGRAPHY

[1] H. C. Chen, *Theory of Electromagnetic Waves: A Coordinate-Free Approach* (McGraw–Hill, New York, NY, 1983). 17

[2] T. G. Mackay and A. Lakhtakia, *The Transfer-Matrix Method in Electromagnetics and Optics* (Morgan & Claypool, San Rafael, CA, 2020). 17, 18, 19, 21, 23, 24, 25, 27

[3] D. Maystre, Ed., *Selected Papers on Diffraction Gratings* (SPIE Optical Engineering Press, Bellingham, WA, 1992). 18

[4] M. G. Moharam, D. A. Pommet, E. B. Grann, and T. K. Gaylord, Stable implementation of the rigorous coupled-wave analysis for surface-relief gratings: Enhanced transmittance matrix approach, *Journal of the Optical Society of America A*, 12:1077–1086 (1995). 18

[5] B. J. Civiletti, A. Lakhtakia, and P. B. Monk, Analysis of the rigorous coupled wave approach for s-polarized light in gratings, *Journal of Computational and Applied Mathematics*, 368:112478 (2020). 18, 24, 28

[6] B. J. Civiletti, A. Lakhtakia, and P. B. Monk, Analysis of the rigorous coupled wave approach for *p*-polarized light in gratings, *Journal of Computational and Applied Mathematics*, 382:113235 (2021). 18, 24, 28

[7] T. Khaleque and R. Magnusson, Light management through guided-mode resonances in thin-film silicon solar cells, *Journal of Nanophotonics*, 8:083995 (2014). 18

[8] M. E. Solano, M. Faryad, A. Lakhtakia, and P. B. Monk, Comparison of rigorous coupled-wave approach and finite element method for photovoltaic devices with periodically corrugated metallic backreflector, *Journal of the Optical Society of America A*, 31:2275–2284 (2014). 18

[9] M. V. Shuba, M. Faryad, M. E. Solano, P. B. Monk, and A. Lakhtakia, Adequacy of the rigorous coupled-wave approach for thin-film silicon solar cells with periodically corrugated metallic backreflectors: Spectral analysis, *Journal of the Optical Society of America A*, 32:1222–1230 (2015). 18

[10] Z. Lokar, B. Lipovsek, M. Topic, and J. Krc, Performance analysis of rigorous coupled-wave analysis and its integration in a coupled modeling approach for optical simulation of complete heterojunction silicon solar cells, *Beilstein Journal of Nanotechnology*, 9:2315–2329 (2018). 18

[11] F. Ahmad, T. H. Anderson, B. J. Civiletti, P. B. Monk, and A. Lakhtakia, On optical-absorption peaks in a nonhomogeneous dielectric material over a two-dimensional metallic surface-relief grating, *Proc. of SPIE*, 10356:103560I (2017). 18

[12] F. Ahmad, T. H. Anderson, B. J. Civiletti, P. B. Monk, and A. Lakhtakia, On optical-absorption peaks in a nonhomogeneous thin-film solar cell with a two-dimensional periodically corrugated metallic backreflector, *Journal of Nanophotonics*, 12:016017 (2018). 18

[13] G. Rajan, K. Aryal, T. Ashrafee, S. Karki, A.-R. Ibdah, V. Ranjan, R. W. Collins, and S. Marsillac, Optimization of anti-reflective coatings for CIGS solar cells via real time spectroscopic ellipsometry, *Proc. of 42nd Photovoltaics Specialist Conference (PVSC)*, New Orleans, LA, June 14–19, 2015. 18

[14] S. R. Kurtz, J. M. Olson, D. J. Friedman, J. F. Geisz, K. A. Bertness, and A. E. Kibbler, Passivation of interfaces in high-efficiency photovoltaic devices, *Materials Research Society Proceedings*, 573:95–106 (1999). 18, 19

[15] B. Vermang, J. T. Wätjen, V. Fjällström, F. Rostvall, M. Edoff, R. Kotipalli, F. Henry, and D. Flandre, Employing Si solar cell technology to increase efficiency of ultra-thin Cu(In,Ga)Se$_2$ solar cells, *Progress in Photovoltaics: Research and Applications*, 22:1023–1029 (2014). 19

[16] M. Solano, M. Faryad, A. S. Hall, T. E. Mallouk, P. B. Monk, and A. Lakhtakia, Optimization of the absorption efficiency of an amorphous-silicon thin-film tandem solar cell backed by a metallic surface-relief grating, *Applied Optics*, 52:966–979 (2013). 19

M. Solano, M. Faryad, A. S. Hall, T. E. Mallouk, P. B. Monk, and A. Lakhtakia, Optimization of the absorption efficiency of an amorphous-silicon thin-film tandem solar cell backed by a metallic surface-relief grating, *Applied Optics*, 54:398–399 (2015) (erratum).

[17] M. Born and E. Wolf, *Principles of Optics*, 6th ed. (Pergamon Press, Oxford, UK, 1980). 19

[18] J. A. Polo Jr., T. G. Mackay, and A. Lakhtakia, *Electromagnetic Surface Waves: A Modern Perspective* (Elsevier, Waltham, MA, 2013). 19, 21, 23, 25

[19] E. N. Glytsis and T. K. Gaylord, Rigorous three-dimensional coupled-wave diffraction analysis of single and cascaded anisotropic gratings, *Journal of the Optical Society of America A*, 4:2061–2080 (1987). 19

[20] H. Lütkepohl, *Handbook of Matrices* (Wiley, Chichester, West Sussex, UK, 1996). 22

[21] Y. Jaluria, *Computer Methods for Engineering* (Taylor and Francis, Washington, DC, 1996). 25

[22] A. Lakhtakia, T. G. Mackay, and C. Zhou, Electromagnetic surface waves at exceptional points, *European Journal of Physics*, 42:015302 (2021). 25

[23] A. Lakhtakia and T. G. Mackay, From unexceptional to doubly exceptional surface waves, *Journal of the Optical Society of America B*, 37:2444–2451 (2020). 25

[24] National Renewable Energy Laboratory, *Reference Solar Spectral Irradiance: Air Mass 1.5* (accessed May 17, 2021). 29

CHAPTER 3

Solar-Cell Electronics

3.1 SEMICONDUCTOR BASICS

3.1.1 CONDUCTORS, SEMICONDUCTORS, AND INSULATORS

The discrete energy states of a material depend upon the electronic constitution of atoms and their bonding strengths [1]. These states are grouped in bands. The highest occupied band, called the valence band, is populated by valence electrons. The lowest unoccupied band is called the conduction band.

In a conductor (usually, a metal), either the valence band is partially occupied or it overlaps with the conduction band so that the availability of empty states at the same energy level makes it easy for valence electrons to be excited and occupy the empty neighboring states in the conduction band. In a semiconductor or an insulator, the valence band is filled and separated from the conduction band by an energy gap called the bandgap energy E_g (usually quantified in eV). This energy gap is wider in insulators than in semiconductors. Hence, insulators have negligibly small conductivity but semiconductors, with E_g ranging from 0.5–3.0 eV, have a small conductivity in the dark. A few valence electrons in a semiconductor have enough energy to move to the conduction band at low temperatures. As the temperature increases, the electrons in the valence band gain some kinetic energy from lattice vibrations and some can even break free. The free electrons are excited into the conduction band and thus contribute to charge transport. The vacancies left behind by those excited electrons are called holes and can also participate in charge transport [2].

A semiconductor can be made to conduct when it is illuminated, for example, by the Sun. Then, the incoming solar photons with energy larger than E_g are absorbed in the semiconductor. The absorbed energy excites electrons in the valence band. The excited electrons move to the conduction band and leave holes behind in the valence band. Thus, electron-hole pairs are created. When an electron and a hole recombine, they cannot contribute to the electric current and their energy is released in thermal and/or electromagnetic forms.

Physical separation of the electrons and holes excited by exposure to light is required to harvest their energy for useful purposes [3, 4]. In a solar cell, the junction of two different extrinsic semiconductors is used to separate the electrons and holes and to harvest them through an external load. This junction is called a *p-n* junction, because one of the two extrinsic semiconductors is of the *p* type as it has an excess of holes and the other is of the *n* type as it has an excess of electrons. The holes (resp., electrons) are in excess when an intrinsic semiconductor is doped by electron-poor (resp., electron-rich) atoms in comparison to the atoms of the intrinsic semiconductor. Some solar cells have a *p-i-n* junction, wherein an intrinsic semiconductor (labeled *i*) separates the two extrinsic semiconductors.

Every semiconductor layer in a solar cell can absorb solar photons with energy equal or greater than E_g. A solar cell's efficiency mainly depends upon the types of semiconductors (i.e., photon-absorbing materials) used therein and their quality. Also, solar-cell performance depends upon the external circuitry, which, however, is not covered in this monograph.

3.1.2   INTRINSIC SEMICONDUCTOR

An intrinsic semiconductor contains small amounts of impurities (or dopants) compared to thermally excited carriers. Let the bandgap energy of an intrinsic semiconductor vary along the z axis. The number of allowed states in the energy range $E(z)$ to $E(z) + dE(z)$ per unit volume is the density of states $N[E(z)]$ and the probability of occupying that energy range is $F[E(z)]$. Thus, the electron density $dn(z)$ in that energy range is the product $N[E(z)] F[E(z)] dE(z)$ [2].

Accordingly, the electron density in the conduction band is given by

$$n(z) \approx \int_{E_c(z)}^{\infty} N[E(z)] F[E(z)] \, dE(z) . \tag{3.1}$$

Likewise, the hole density in the valence band is given by

$$p(z) \approx \int_{-\infty}^{E_v(z)} N[E(z)] \{1 - F[E(z)]\} \, dE(z) , \tag{3.2}$$

with $1 - F[E(z)]$ as the probability of states in the valence band being unoccupied. Whereas $E_v(z)$ is the highest energy in the valence band, $E_c(z)$ is the lowest energy in the conduction band, so that the bandgap energy

$$E_g(z) = E_c(z) - E_v(z) . \tag{3.3}$$

As the probability of occupation obeys the Fermi–Dirac statistics [2],

$$F[E(z)] = \frac{1}{1 + \exp\{[E(z) - E_F(z)]/k_B T\}} \tag{3.4}$$

is the Fermi–Dirac distribution function, where $k_B = 1.3806 \times 10^{-23}$ J K^{-1} is the Boltzmann constant, T is the absolute temperature, and $E_F(z)$ is the Fermi level. The Fermi level is defined as the energy of the highest-energy state that is notionally occupied at absolute zero temperature.

In the Boltzmann approximation, the conduction band-edge energy $E_c(z)$ exceeds the Fermi level $E_F(z)$ by $\geq 3k_B T$ and the valence band-edge energy $E_v(z)$ is $\geq 3k_B T$ below $E_F(z)$; hence, the Fermi–Dirac distribution function $F[E(z)]$ can be approximated by the Maxwell–Boltzmann distribution [3]

$$F_{\mathrm{MB}}[E(z)] = \exp\{-[E(z) - E_F(z)]/k_B T\} . \tag{3.5}$$

The integrals on the right sides of Eqs. (3.1) and (3.2) then deliver

$$n(z) = N_c(z) \exp\left\{ [E_F(z) - E_c(z)]/k_BT \right\} \tag{3.6}$$

and

$$p(z) = N_v(z) \exp\left\{ [E_v(z) - E_F(z)]/k_BT \right\}, \tag{3.7}$$

where $N_c(z)$ is the effective density of states in the conduction band and $N_v(z)$ is the effective density of states in the valence band.

According to the law of mass action, the product of $n(z)$ and $p(z)$ is constant under thermal equilibrium, whereby the intrinsic carrier density

$$n_i(z) = \sqrt{n(z)\, p(z)} \tag{3.8}$$

is defined. Therefore, Eqs. (3.6)–(3.8) yield

$$n_i(z) = \sqrt{N_c(z)N_v(z)}\, \exp\left[-E_g(z)/2k_BT \right]. \tag{3.9}$$

Equations (3.6) and (3.7) can then be rewritten as

$$n(z) = n_i(z) \exp\left\{ [E_F(z) - E_i(z)]/k_BT \right\} \tag{3.10}$$

and

$$p(z) = n_i(z) \exp\left\{ [E_i(z) - E_F(z)]/k_BT \right\}, \tag{3.11}$$

respectively, where

$$E_i(z) = \frac{1}{2}\left[E_c(z) + E_v(z) \right] - \frac{1}{2}k_BT \ln\left[\frac{N_c(z)}{N_v(z)} \right] \tag{3.12}$$

is the intrinsic energy of charge carriers. According to this definition, $E_i(z)$ lies close to the center of the bandgap and it is equal to the Fermi level of the intrinsic semiconductor.

An important quantity is the electron affinity $\chi(z)$ of the semiconductor. It is the least amount of energy required to remove an electron from the material. It is measured from the lowest energy in the conduction band (i.e., $E_c(z)$) to the vacuum energy $E_{vac}(z)$, which is the energy required to free the electron from all forces in the material. Thus [3],

$$E_c(z) = E_{vac}(z) - \chi(z) \tag{3.13}$$

so that

$$E_v(z) = E_{vac}(z) - \chi(z) - E_g(z) \tag{3.14}$$

follows from Eq. (3.3). The intrinsic energy $E_i(z)$ can be related to $E_{vac}(z)$ and $\chi(z)$ by using Eqs. (3.13) and (3.14) in Eq. (3.12).

The vacuum energy $E_{vac}(z)$ is related to an electrostatic potential $\phi(z)$ as

$$E_{vac}(z) = E_0 - q_e\phi(z), \tag{3.15}$$

where $q_e = 1.602 \times 10^{-19}$ C and the arbitrary reference energy E_0 is usually chosen as follows:

$$E_0 = E_c(0) + q_e\phi(0) + \chi(0). \tag{3.16}$$

3.1.3   EXTRINSIC SEMICONDUCTOR

The properties of an intrinsic semiconductor can be altered by doping it with impurities. Doped semiconductors are also called extrinsic semiconductors. An n-type semiconductor has electrons as the majority charge carriers because it is doped with electron-rich impurities (or defects) that can donate electrons, the donor density or concentration being denoted by $N_D(z)$. Similarly, a p-type semiconductor has holes as the majority charge carriers because it contains electron-poor impurities (or defects) that can accept electrons and, equivalently, donate holes; $N_A(z)$ denotes the acceptor density or concentration.

In an n-type semiconductor, all the surplus electrons are thermally excited and free to move to the conduction band at room temperature and $N_D(z) \gg n_i(z)$; hence, the concentration of conduction electrons is approximately equal to that of impurities: $n(z) \approx N_D(z)$. Since the semiconductor is still at equilibrium, Eq. (3.3) holds so that $p(z) = n_i^2(z)/N_D(z)$ follows from Eq. (3.8). The electrons are the majority carriers, whereas the holes are the minority carriers. The Fermi level in terms of $N_c(z)$ and $N_D(z)$ becomes [3]

$$E_F(z) = E_c(z) + k_B T \, \ln\left[N_D(z)/N_c(z)\right] . \tag{3.17}$$

Similarly, in a p-type semiconductor, the density of valence holes is approximately equal to the density of acceptor impurities; hence, $p(z) \approx N_A(z)$ and $n(z) = n_i^2(z)/N_A(z)$. The corresponding Fermi level in terms of $N_v(z)$ and $N_A(z)$ is given by

$$E_F(z) = E_v(z) - k_B T \, \ln\left[N_A(z)/N_v(z)\right] . \tag{3.18}$$

In a semiconductor that is either intentionally doped or has intrinsic defects, Eq. (3.3) still holds, but with the additional charge neutrality condition

$$p(z) + N_D(z) = n(z) + N_A(z) . \tag{3.19}$$

3.1.4   SEMICONDUCTOR UNDER EXTERNAL PERTURBATION

So far, the semiconductor was considered to be in equilibrium, i.e., no external perturbation was assumed to have been applied. When a solar cell is perturbed by sunlight or external bias, it is not under equilibrium anymore. Equations (3.10) and (3.11) then do not hold and $n_i^2(z) \neq n(z)p(z)$.

However, if the perturbation is not large or does not change very quickly, the electron and hole densities can be calculated by assuming that the populations of electrons and holes relax to achieve a state of quasi-thermal equilibrium such that the electrons within the conduction band distribute themselves as if they are at equilibrium with a common Fermi level, and the holes distribute themselves within the valence band as if they are at equilibrium with a common Fermi level. The quasi-Fermi levels for electrons and holes are denoted by $E_{F_n}(z)$ and $E_{F_p}(z)$, respectively. Then,

$$n(z) = n_i(z) \exp\left\{\left[E_{F_n}(z) - E_i(z)\right]/k_B T\right\} \tag{3.20}$$

and

$$p(z) = n_i(z) \exp \left\{ [\mathsf{E}_i(z) - \mathsf{E}_{F_p}(z)]/k_B T \right\} \tag{3.21}$$

replace Eqs. (3.10) and (3.11), respectively. The electron quasi-Fermi level is given by

$$\mathsf{E}_{F_n}(z) = \mathsf{E}_c(z) + k_B T \ln [n(z)/N_c(z)] \tag{3.22}$$

and the hole quasi-Fermi level by

$$\mathsf{E}_{F_p}(z) = \mathsf{E}_v(z) - k_B T \ln [p(z)/N_v(z)] . \tag{3.23}$$

In terms of conduction and valence band-edge energies, the electron and hole densities become

$$n(z) = N_c(z) \exp \left\{ [\mathsf{E}_{F_n}(z) - \mathsf{E}_c(z)]/k_B T \right\} \tag{3.24}$$

and

$$p(z) = N_v(z) \exp \left\{ [\mathsf{E}_v(z) - \mathsf{E}_{F_p}(z)]/k_B T \right\} , \tag{3.25}$$

respectively.

3.1.5   CURRENT DENSITIES UNDER EXTERNAL PERTURBATION

The electron-current density $J_n(z)$ and the hole-current density $J_p(z)$ under external perturbation are driven by gradients in the electron and hole quasi-Fermi levels, respectively [2, 3]. Thus,

$$J_n(z) = \mu_n(z) \, n(z) \frac{d}{dz} \mathsf{E}_{F_n}(z) \tag{3.26a}$$

and

$$J_p(z) = \mu_p(z) \, p(z) \frac{d}{dz} \mathsf{E}_{F_p}(z), \tag{3.26b}$$

where $\mu_n(z)$ is the electron mobility and $\mu_p(z)$ is the hole mobility. After using Eqs. (3.22) and (3.23), $J_n(z)$ and $J_p(z)$ can be simplified to

$$J_n(z) = -q_e \mu_n(z) \left\{ n(z) \frac{d}{dz} [\phi(z) + \phi_n(z)] - V_{th} \frac{d}{dz} n(z) \right\} \tag{3.27a}$$

and

$$J_p(z) = -q_e \mu_p(z) \left\{ p(z) \frac{d}{dz} [\phi(z) + \phi_p(z)] + V_{th} \frac{d}{dz} p(z) \right\} , \tag{3.27b}$$

respectively. Here,

$$\phi_n(z) = \left\{ \chi(z)/q_e + V_{th} \ln \left[\frac{N_c(z)}{N_0} \right] \right\} \tag{3.28}$$

and

$$\phi_p(z) = \left\{ \chi(z)/q_e + \mathsf{E}_g(z)/q_e - V_{th} \ln \left[\frac{N_v(z)}{N_0} \right] \right\} \tag{3.29}$$

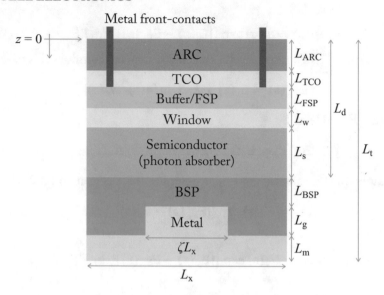

Figure 3.1: Schematic of the unit cell of a thin-film solar cell backed by a 1D periodically corrugated back reflector.

are the built-in potentials for electrons and holes (due to variations in the material properties), respectively, and

$$V_{th} = k_B T/q_e \tag{3.30}$$

is the thermal voltage. Both built-in potentials as well as the electron affinity may be discontinuous with respect to position at heterojunctions in the semiconductor region. The baseline number density N_0 is arbitrary because potentials are defined uniquely only up to a constant. In Eqs. (3.27a) and (3.27b), the contribution of diffusion of electrons to $J_n(z)$ can be identified as depending on $dn(z)/dz$, the remainder being the contribution of electron drift, and likewise for $J_p(z)$.

3.2   TRANSPORT IN SOLAR CELLS

The solar-cell structure introduced in Chapter 2 for optical analysis is reproduced in Fig. 3.1 for electrical analysis. The TCO layer is connected to the front electrode and the metal back reflector is the back electrode. Thus, electrical analysis involves the region $L_{ARC} + L_{TCO} < z < L_d + L_{BSP}$. The buffer or the front-surface passivation layer, the window layer, and the photon-absorbing layer are made of semiconductors and contribute to charge-carrier generation. If the back-surface passivation layer is made of a semiconductor, it also contributes to charge-carrier generation, but if that layer is made of an insulator, it does not. Thus, the region $\mathcal{R}_{sc}$ comprises either three layers or four. The back-surface passivation layer is thin enough that it can be neglected for electrical analysis, if the material is not a semiconductor. As the anti-reflection coating and the back reflector do not contribute toward the electron-hole-pair generation rate, they need not be included for electrical analysis. However, the optical effects of these layers are included in the optical analysis described in Chapter 2. As the focus in this monograph is on modeling the optoelectronic characteristics of the solar cell, not on how it interfaces with an external circuit, the planes $z = L_{ARC} + L_{TCO}$ and $z = L_d + L_{BSP}$ are assumed to be ideal ohmic contacts.

A 1D drift-diffusion model [3–5] used to investigate the transport of electrons and holes for $z \in [L_{ARC} + L_{TCO}, L_d]$ is discussed in the remainder of this chapter with the assumption that $L_{BSP} = 0$.

3.2.1 1D ELECTRON-HOLE-PAIR GENERATION RATE

The optical domain $0 < z < L_t$ analyzed in Chapter 2 is used to compute the generation rate $G(x, z)$ in $\mathcal{R}_{sc}$. As the solar cell operates under the influence of a z-directed electrostatic field created by the application of a bias voltage V_{ext}, charge carriers generally flow along the z axis; any current generated parallel to the x axis is very small. Moreover, the period L_x of the periodically corrugated back reflector is ∼500 nm, which is so small in comparison to the lateral dimensions of the solar cell that it can be ignored for electrostatic analysis. Therefore, the x-averaged electron-hole-pair generation rate is calculated as

$$G(z) = \frac{1}{L_x} \int_{-L_x/2}^{L_x/2} G(x, z) \, dx, \qquad z \in (L_{ARC} + L_{TCO}, L_d) . \tag{3.31}$$

The generation rate $G(z)$ contains the effects of the periodic corrugations of the back reflector, both passivation layers, and the anti-reflection coating.

3.2.2 1D DRIFT-DIFFUSION MODEL

The number of charge carriers of each polarity must be conserved. Furthermore, the electrostatic potential $\phi(z)$ due to all charge carriers must obey the Poisson equation. Under steady-state conditions, the 1D drift-diffusion model comprises the following three differential equations [3]:

$$\frac{d}{dz} J_n(z) = -q_e \left[G(z) - R(n, p; z) \right] , \tag{3.32a}$$

$$\frac{d}{dz} J_p(z) = q_e \left[G(z) - R(n, p; z) \right] , \tag{3.32b}$$

and

$$\varepsilon_0 \frac{d}{dz} \left[\varepsilon_{dc}(z) \frac{d}{dz} \phi(z) \right] = -q_e \left[N_f(z) + N_d(z) + p(z) - n(z) \right] . \tag{3.32c}$$

These three differential equations hold for $z \in (L_{ARC} + L_{TCO}, L_d)$, with $N_f(z)$ as the density (also called trap density) due to intrinsic defects (also called traps), $N_d(z) = N_D(z) - N_A(z)$ as the excess of the donor density over the acceptor density, $R(n, p; z)$ as the electron-hole-pair recombination rate, and $\varepsilon_{dc}(z)$ as the dc relative permittivity.

Equations (3.27) and (3.32) have to be solved concurrently for $z \in (L_{ARC} + L_{TCO}, L_d)$, in conjunction with a set of boundary conditions for $n(z)$, $p(z)$, and $\phi(z)$ on the planes $z = L_{ARC} + L_{TCO}$ and $z = L_d$ [6]. To model both planes as ideal ohmic contacts, the Dirichlet boundary conditions [3, 4]

$$n(L_{ARC} + L_{TCO}) = n_0(L_{ARC} + L_{TCO}), \tag{3.33a}$$

$$p(L_{ARC} + L_{TCO}) = p_0(L_{ARC} + L_{TCO}), \tag{3.33b}$$

$$\phi(L_{ARC} + L_{TCO}) = \phi_0(L_{ARC} + L_{TCO}) + V_{ext}, \tag{3.33c}$$

$$n(L_d) = n_0(L_d), \tag{3.33d}$$

$$p(L_d) = p_0(L_d), \tag{3.33e}$$

and

$$\phi(L_d) = \phi_0(L_d) \tag{3.33f}$$

are used. Herein, $n_0(z)$, $p_0(z)$, and $\phi_0(z)$ are the electron density, hole density and potential at local quasi-thermal equilibrium, respectively, as discussed in Section 3.2.3, while V_{ext} is the externally applied (i.e., bias) voltage across the two planes. Solution of the system of five equations (3.27) and (3.32), subject to the six boundary conditions (3.33), enables the calculation of the current density

$$J(z) = J_n(z) + J_p(z) \tag{3.34}$$

in the solar cell.

3.2.3   LOCAL QUASI-THERMAL EQUILIBRIUM

If a region containing a homogeneous semiconductor is charge-free and isolated from any external influences (e.g., $G \equiv 0$ and $V_{ext} = 0$), the distributions of electrons and holes for energy tend to the Fermi–Dirac distribution, with the Fermi level lying at the intrinsic energy level, i.e., $E_{F_n} = E_{F_p} = E_F = E_i$. Then, $J_n = J_p$ is identically zero, and the semiconductor is said to be in local quasi-thermal equilibrium, i.e.,

$$n_i^2(z) = n_0(z)\, p_0(z) \tag{3.35a}$$

and

$$N_f(z) + N_d(z) + p_0(z) - n_0(z) = 0 \tag{3.35b}$$

for all $z \in (L_{ARC} + L_{TCO}, L_d)$.

This concept is used to model the ideal ohmic boundary conditions mentioned in Section 3.2.2. As shown by Anderson et al. [6],

$$n_0(z) = \frac{[N_f(z) + N_d(z)] + \sqrt{[N_f(z) + N_d(z)]^2 + 4n_i^2(z)}}{2} ,$$
(3.36)

$$p_0(z) = \frac{-[N_f(z) + N_d(z)] + \sqrt{[N_f(z) + N_d(z)]^2 + 4n_i^2(z)}}{2} ,$$
(3.37)

and

$$\phi_0(z) = -\frac{1}{q_e} \left\{ [\chi(z) - \chi(0)] - \frac{1}{2}[E_g(z) - E_g(0)] \right\}$$
$$+ \frac{1}{2}V_{th} \ln \left[\frac{n_0(z)}{n_0(0)} \right] + \frac{1}{2}V_{th} \ln \left[\frac{p_0(z)}{p_0(0)} \right]$$
$$- \frac{1}{2}V_{th} \ln \left[\frac{N_c(z)}{N_c(0)} \right] - \frac{1}{2}V_{th} \ln \left[\frac{N_v(z)}{N_v(0)} \right] .$$
(3.38)

Thus, expressions for the right sides of Eqs. (3.33) can be formulated. Note, however, that $n(z) \neq n_0(z)$, $p(z) \neq p_0(z)$, and $\phi(z) \neq \phi_0(z)$ for $z \in (L_{ARC} + L_{TCO}, L_d)$ in general because the right side of Eq. (3.32c) cannot be identically zero but the right side of Eq. (3.35b) is.

3.2.4   RECOMBINATION PROCESSES

Recombination of an electron and a hole can take place through several different mechanisms [3, 4]. The three main ones are named: radiative, Shockley–Read–Hall (SRH), and Auger.

Radiative Recombination

Radiative recombination occurs when an electron and a hole recombine across the full bandgap, releasing the energy as a photon with energy equal to the bandgap energy. At quasi-thermal equilibrium, radiative recombination is identically balanced by electrons being thermally excited across the bandgap. The radiative recombination rate is given by [3, 6]

$$R_{rad}(n, p; z) = R_B(z) \left[n(z)p(z) - n_i^2(z) \right] ,$$
(3.39)

where R_B is the radiative recombination coefficient. It should be noted that $R_{rad}(n, p; z) \equiv 0 \; \forall z \in [L_{ARC} + L_{TCO}, L_d]$ at local quasi-thermal equilibrium.

Shockley–Read–Hall Recombination

The SRH recombination mechanism involves electrons and holes recombining via an intermediate-gap state. It is modeled by [3, 6]

$$R_{SRH}(n, p; z) = \frac{n(z)p(z) - n_i^2(z)}{\tau_p(z) [n(z) + n_1(z)] + \tau_n(z) [p(z) + p_1(z)]} ,$$
(3.40)

where

$$n_1(z) = n_i(z) \exp[(E_T - E_i)/k_B T] \qquad (3.41a)$$

and

$$p_1(z) = n_i(z) \exp[(E_i - E_T)/k_B T] \qquad (3.41b)$$

are the electron and hole densities at the trap energy level E_T, respectively. If this level is the intrinsic energy level $E_i(z)$, then $n_1(z) = p_1(z) = n_i(z)$ from Eqs. (3.20) and (3.21). The functions $\tau_n(z)$ and $\tau_p(z)$ are the minority carrier lifetimes given as

$$\tau_n(z) = \frac{1}{\sigma_n(z) v_{th,n}(z) N_f(z)} \qquad (3.42a)$$

and

$$\tau_p(z) = \frac{1}{\sigma_p(z) v_{th,p}(z) N_f(z)} , \qquad (3.42b)$$

respectively, where $\sigma_n(z)$ is the capture cross section for electrons, $\sigma_n(z)$ is the capture cross section for holes, $v_{th,n}(z)$ is the mean thermal speed of electrons, and $v_{th,p}(z)$ is the mean thermal speed of holes.

Auger Recombination

The Auger recombination mechanism involves a three-particle recombination pathway, occurring when an electron and a hole recombine across the bandgap, with the released energy transferred to a third charge carrier which is excited away from the edges of the valence and conduction bands. This recombination rate is given by [3, 6]

$$\begin{aligned} R_{Aug}(n, p; z) = &\ C_n(z) n(z) \left[n(z) p(z) - n_i^2(z) \right] \\ &+ C_p(z) p(z) \left[n(z) p(z) - n_i^2(z) \right] , \end{aligned} \qquad (3.43)$$

where the functions $C_n(z)$ and $C_p(z)$ are Auger recombination coefficients.

All three contributions add, so that

$$R(n, p; z) = R_{rad}(n, p; z) + R_{SRH}(n, p; z) + R_{Aug}(n, p; z) \qquad (3.44)$$

is the total recombination rate. Note that $R(n, p; z) = 0$ at local quasi-thermal equilibrium.

3.2.5   HETEROJUNCTIONS: CONTINUOUS QUASI-FERMI LEVELS

When there is a jump in either $\chi(z)$ or $E_g(z)$ or both, and thus in $\phi_n(z)$ or $\phi_p(z)$, then $n(z)$ and $p(z)$ may also have jumps. A heterojunction is formed at the discontinuity [7]. Commonly used methods to model this discontinuity invoke the assumption of either continuous quasi-Fermi levels or thermionic emission [8].

Continuous quasi-Fermi levels are well suited for modeling the heterojunction discontinuity [6]. To quantify the jump in $n(z)$, $J_n(z)$ in Eq. (3.27a) is rewritten as

$$J_n(z) = q_e \mu_n(z) V_{th} \frac{d}{dz} \left\{ n(z) \exp\left[-\frac{\phi(z) + \phi_n(z)}{V_{th}} \right] \right\} \exp\left[\frac{\phi(z) + \phi_n(z)}{V_{th}} \right]. \qquad (3.45)$$

Since $n(z)$, $\phi(z)$, and $\phi_n(z)$ are smooth functions of z away from a heterojunction, we can avoid the occurrence of the Dirac delta in the expression for $J_n(z)$ by requiring that

$$n(z) \exp\left[-\frac{\phi(z) + \phi_n(z)}{V_{th}} \right]$$

is continuous across a heterojunction. Accordingly, if the heterojunction occurs at $z = z_H$,

$$n(z_H^-) \exp\left[-\frac{\phi(z_H^-) + \phi_n(z_H^-)}{V_{th}} \right] = n(z_H^+) \exp\left[-\frac{\phi(z_H^+) + \phi_n(z_H^+)}{V_{th}} \right], \qquad (3.46)$$

where $n(z_H^-)$ is the limiting value of $n(z_H - \delta_z)$ as the non-negative δ_z goes to 0, $n(z_H^+)$ is the limiting value of $n(z_H + \delta_z)$ as $\delta_z \to 0$, and $\phi(z_H^\pm)$ as well as $\phi_n(z_H^\pm)$ are similarly defined. Since $\phi(z)$ is continuous across the plane $z = z_H$ because it is the solution of a Poisson problem, Eq. (3.46) yields

$$n(z_H^-) = n(z_H^+) \exp\left[\frac{\phi_n(z_H^-) - \phi_n(z_H^+)}{V_{th}} \right]. \qquad (3.47)$$

A similar argument yields

$$p(z_H^-) = p(z_H^+) \exp\left[-\frac{\phi_p(z_H^-) - \phi_p(z_H^+)}{V_{th}} \right]. \qquad (3.48)$$

Alternative derivations of Eqs. (3.47) and (3.48) are available in Section 3.4 of Ref. 6.

3.3   PERFORMANCE PARAMETERS

Under steady-state conditions, $J(z) = J_{dev}$ is uniform between the two ideal ohmic contacts. Thus, J_{dev} is the current density delivered to an external circuit, J_{sc} is the value of J_{dev} when $V_{ext} = 0$, and V_{oc} is the value of V_{ext} such that $J_{dev} = 0$. The device current density J_{dev} depends on the choice of V_{ext}, i.e., $J_{dev} \equiv J_{dev}(V_{ext})$. Repeating calculations for various values of V_{ext} produces the J_{dev}-V_{ext} curve. Likewise, the power density as a function of V_{ext} can be calculated as

$$P(V_{ext}) = J_{dev}(V_{ext}) V_{ext} . \qquad (3.49)$$

The maximum power density obtainable from the solar cell is

$$P_{max} = \max_{V_{ext}} P(V_{ext}) . \qquad (3.50)$$

The power-conversion efficiency of the solar cell is calculated as

$$\eta = \frac{P_{\text{max}}}{P_{\text{in}}},$$ (3.51)

where $P_{\text{in}} = 1000 \text{ W m}^{-2}$ is the incident solar power density. It is the efficiency of the solar cell that must be optimized. Another performance parameter for solar cell is the fill factor

$$\text{FF} = \frac{P_{\text{max}}}{J_{\text{sc}} V_{\text{oc}}}.$$ (3.52)

3.4   NUMERICAL APPROXIMATION OF THE DRIFT DIFFUSION SYSTEM

After the generation rate $G(z)$ due to solar illumination has been computed using Eq. (3.31), the drift-diffusion system given by Eqs. (3.32) with the boundary conditions (3.33) needs to be solved for various values of V_{ext} in order to predict the power-conversion efficiency η of the solar cell. This is conventionally done using the Sharfetter–Gummel discretization method [9], which is an exponentially upwinded finite-difference method designed to handle the regions wherein convection is dominant as well as regions in which diffusion dominates. The importance of the drift-diffusion system has led to the investigation of a wide variety of other finite-difference and finite-element methods; see, e.g., Refs. 10 and 11.

The hybridizable discontinuous Galerkin (HDG) method has recently emerged for simulating organic solar cells [12]. The HDG method has also been successfully applied to pure diffusion problems [13], convection-dominated problems [14], as well as more complex fluid-flow simulations [15]. It combines the benefits of higher-order accuracy and, in higher dimensions, the ability to model complex geometry. In addition, it has the simple local approximation of a finite-volume method. Hence, it often performs better than both the finite-element and finite-volume methods [17].

The HDG method possesses several other features that are advantageous for the solution of the drift-diffusion system for thin-film solar cells [12]. The relaxation of the requirement that the approximate solution be continuous at the boundaries between elements permits solutions where the variables have strong gradients and second derivatives. It also naturally allows discontinuities in the solution which may arise due to discontinuous material parameters, such as those which occur at heterojunctions. Therefore, this method is very suitable for the electrical submodel.

3.4.1   HYBRIDIZABLE DISCONTINUOUS GALERKIN METHOD

Application of the HDG method to the drift-diffusion system requires subdivision of the region $L_{\text{ARC}} + L_{\text{TCO}} \leq z \leq L_{\text{d}}$ into a mesh of geometric elements using $N_{\text{z}} + 1$ grid points $\{z_{\gamma}^{h}\}_{\gamma=0}^{N_{\text{z}}}$

satisfying $L_{\text{ARC}} + L_{\text{TCO}} = z_0^h < z_1^h < \cdots < z_{N_z}^h = L_d$, with the mesh parameter

$$h = \max_{1 \leq \gamma \leq N_z} |z_\gamma^h - z_{\gamma-1}^h|. \tag{3.53}$$

The mesh is chosen such that any discontinuities in the material properties (i.e., at boundaries of two different materials) occur at a grid point. The geometric elements are then the intervals $K_\gamma^h = (z_{\gamma-1}^h, z_\gamma^h)$, $\gamma \in [1, N_z]$, with $\mathcal{T}_h$ denoting the collection of all these intervals.

The drift-diffusion equations (3.32) now have to be discretized [6]. As an example, consider the electrostatic potential $\phi(z)$ that satisfies Eq. (3.32c) along with the Dirichlet boundary conditions (3.33c) and (3.33f) [13]. Since the HDG method is usually applied to a first-order system, the electrostatic displacement $D_{\text{dc}}(z)\hat{u}_z$ is introduced to obtain

$$\frac{d}{dz} D_{\text{dc}}(z) = q_e [N_f(z) + N_d(z) + p(z) - n(z)] \tag{3.54a}$$

from Eq. (3.32c), where

$$D_{\text{dc}}(z) = -\varepsilon_0 \varepsilon_{\text{dc}}(z) \frac{d}{dz} \phi(z). \tag{3.54b}$$

Both sides of Eq. (3.54a) are multiplied by a smooth test function Ψ_γ and integrated with respect to z over K_γ^h; likewise, both sides of Eq. (3.54b) are multiplied by a test function ψ_γ followed by integration over K_γ^h. Integration by parts is used to transfer derivatives to the test functions, so that the equations

$$\int_{K_\gamma^h} D_{\text{dc}}(z) \frac{d}{dz} \Psi_\gamma(z)\, dz - \left[D_{\text{dc}}(z_\gamma^h) \Psi_\gamma(z_\gamma^h) - D_{\text{dc}}(z_{\gamma-1}^h) \Psi_\gamma(z_{\gamma-1}^h) \right]$$

$$+ q_e \int_{K_\gamma^h} [p(z) - n(z)]\, \Psi_\gamma(z)\, dz = -q_e \int_{K_\gamma^h} [N_f(z) + N_d(z)]\, \Psi_\gamma(z)\, dz \tag{3.55a}$$

and

$$\int_{K_\gamma^h} \phi(z) \frac{d}{dz} \psi_\gamma^h(z)\, dz - \left[\phi(z_\gamma^h)\psi_\gamma(z_\gamma^h) - \phi(z_{\gamma-1}^h)\psi_\gamma(z_{\gamma-1}^h) \right]$$

$$- \frac{1}{\varepsilon_0} \int_{K_\gamma^h} \frac{1}{\varepsilon_{\text{dc}}(z)} D_{\text{dc}}(z)\psi_\gamma(z)\, dz = 0 \tag{3.55b}$$

result.

Let equal-degree piecewise-polynomial spaces be used to discretize both ϕ and D_{dc} in Eqs. (3.55) [13]. In particular, let the discontinuous finite-element space

$$\mathbb{V}_h = \left\{ v_h \in L^2(\Omega_{\text{sc}}) \,|\, v_h \mid_{K_\gamma^h} \in \mathbb{P}_{P_{\text{deg}}} \text{ for all } K_\gamma^h \in \mathcal{T}_h \right\} \tag{3.56}$$

be used, where the domain $\Omega_{\text{sc}} \equiv \{z \in (L_{\text{ARC}} + L_{\text{TCO}}, L_{\text{d}})\}$ and $\mathbb{P}_{P_{\text{deg}}}$ is the set of polynomials in one variable of maximum degree $P_{\text{deg}} > 0$. In each element K_{γ}^h, the $P_{\text{deg}} + 1$ Gauss–Lobatto quadrature points $\{z_{\gamma,p}^{GL,h}\}_{p=0}^{P_{\text{deg}}}$ are used as the interpolation points of the finite-element functions. In particular, the shape functions $\psi_{\gamma,q}^h \in \mathbb{P}_{P_{\text{deg}}}$ are defined for $0 \le q \le P_{\text{deg}}$ by requiring that $\psi_{\gamma,q}^h(z_{\gamma,p}^{GL,h}) = \delta_{pq}$ for $0 \le p \le P_{\text{deg}}$ and $\psi_{\gamma,q}^h(z) = 0$ for $z \notin K_{\gamma}^h$. These functions allow us to expand any function $v_h \in \mathbb{V}_h$ as

$$v_h(z) = \sum_{\gamma=1}^{N_z} \sum_{p=0}^{P_{\text{deg}}} v_{\gamma,p}^h \, \psi_{\gamma,p}^h(z) \tag{3.57}$$

with suitable coefficients $v_{\gamma,p}^h$. Equation (3.57) is used to compute certain matrixes involved in the HDG method.

For any two functions $v(z)$ and $w(z)$ with $v|_{K_{\gamma}^h} \in L^2(K_{\gamma}^h)$ and $w|_{K_{\gamma}^h} \in L^2(K_{\gamma}^h)$ for $1 \le \gamma \le N_z$, the inner product

$$(v, w)_{\mathcal{T}_h} = \sum_{\gamma=1}^{N_z} \int_{K_{\gamma}^h} v(z)w(z)\, dz \tag{3.58}$$

is defined. Also, for any function $v(z)$ that is piecewise smooth on the mesh, the left and right limits at a grid point z_{γ}^h are defined as

$$v(z_{\gamma}^{h,-}) = \lim_{\substack{z \to z_{\gamma}^h \\ z < z_{\gamma}^h}} v(z) \tag{3.59a}$$

and

$$v(z_{\gamma}^{h,+}) = \lim_{\substack{z \to z_{\gamma}^h \\ z > z_{\gamma}^h}} v(z), \tag{3.59b}$$

respectively. Finally, with $\partial \mathcal{T}_h$ denoting the set of endpoints of the mesh intervals, for any $\hat{v} \in \mathbb{R}^{N_z+1}$ and $\Psi_h \in \mathbb{V}_h$,

$$\langle \hat{v}, \Psi_h \rangle_{\partial \mathcal{T}_h} = \sum_{\gamma=1}^{N_z} \left[\hat{v}_{\gamma} \Psi_h(z_{\gamma}^{h,-}) - \hat{v}_{\gamma-1} \Psi_h(z_{\gamma-1}^{h,+}) \right]. \tag{3.60}$$

Next, the function D_{dc} is replaced by $D_h \in \mathbb{V}_h$, n by $n_h \in \mathbb{V}_h$, p by $p_h \in \mathbb{V}_h$, and ϕ by $\phi_h \in \mathbb{V}_h$ in Eqs. (3.55), and the test functions therein are chosen to be finite-element functions $\psi \in \mathbb{V}_h$ and $\Psi \in \mathbb{V}_h$. Because the finite-element functions in $\mathbb{V}_h$ are discontinuous between elements, the grid-point values of ϕ_h and D_h on each element in Eq. (3.54b) are no longer well defined; this is indicated by a ^ over the relevant quantities, and we shall shortly detail how these terms are handled. We obtain

$$\left(D_h, \frac{d}{dz}\Psi_h\right)_{\mathcal{T}_h} - \langle\hat{D}_h, \Psi_h\rangle_{\partial\mathcal{T}_h} + q_h(p_h - n_h, \Psi_h)_{\partial\mathcal{T}_h} = -q_e(N_f + N_d, \Psi_h)_{\mathcal{T}_h} \qquad (3.61a)$$

for all $\Psi \in \mathbb{V}_h$ and

$$\left(\phi_h, \frac{d}{dz}\psi\right)_{\mathcal{T}_h} - \langle\hat{\phi}_h, \psi_h\rangle_{\partial\mathcal{T}_h} - \left(\frac{1}{\varepsilon_0\varepsilon_{dc}}D_h, \psi\right)_{\mathcal{T}_h} = 0 \qquad (3.61b)$$

for all $\psi \in \mathbb{V}_h$.

After making the standard choice that $\hat{\phi}_h$ should be determined by the HDG method, we choose $\hat{\phi}_h \in \mathbb{R}^{N_z+1}$. Furthermore, $\hat{D}_h$ is defined in two steps as follows. On every element K_γ^h, we define two *numerical fluxes* as

$$\hat{D}_\gamma^- = D_h(z_\gamma^{h-}) + \tau_\gamma^- \left[\phi_h(z_\gamma^{h,-}) - \hat{\phi}_{h,\gamma}\right] \qquad (3.62a)$$

and

$$\hat{D}_{\gamma-1}^+ = D_h(z_{\gamma-1}^{h+}) + \tau_{\gamma-1}^+ \left[\phi_h(z_{\gamma-1}^{h,+}) - \hat{\phi}_{h,\gamma-1}\right], \qquad (3.62b)$$

with the positive penalty parameters τ_γ^- and $\tau_{\gamma-1}^+$ usually taken to be $O(1)$ independent of h. The continuity of the numerical flux for D_h is enforced by requiring that

$$D_\gamma^- = D_\gamma^+, \quad \gamma \in [1, N_z - 1]. \qquad (3.63)$$

Finally, the boundary conditions are enforced by requiring that

$$\hat{\phi}_{h,0} = \phi_0(L_{ARC} + L_{TCO}) + V_{ext} \qquad (3.64a)$$

and

$$\hat{\phi}_{h,N_z} = \phi_0(L_d). \qquad (3.64b)$$

Taken together, Eqs. (3.61)–(3.64) define $2N_z(P_{deg} + 1) + (N_z + 1)$ linear equations coupling the $2N_z(P_{deg} + 1) + (N_z + 1)$ degrees of freedom for $(\phi_h, \hat{\phi}_h, D_h) \in \mathbb{V}_h \times \mathbb{R}^{N_z+1} \times \mathbb{V}_h$ to the $2N_z(P_{deg} + 1)$ degrees of freedom for $n_h \in \mathbb{V}_h$ and $p_h \in \mathbb{V}_h$.

The foregoing procedure to handle Eqs. (3.32c), (3.33c), and (3.33f) is also applied to [6]:

- Eqs. (3.27a), (3.32a), (3.33a), and (3.33d) to calculate $J_n(z)$; and

- Eqs. (3.27b), (3.32b), (3.33b), and (3.33e) to calculate $J_p(z)$.

The finite-element functions $J_{n,h} \in \mathbb{V}_h$ and $J_{p,h} \in \mathbb{V}_h$ are postulated as the respective approximations of J_n and J_p. The following three complications occur. (i) The right sides of Eqs. (3.32a) and (3.32b) are nonlinear due to the recombination rate and needs to be handled efficiently. (ii) The penalty parameters in the numerical fluxes need to be defined in very specific ways, depending on charge-carrier drift. (iii) As the charge densities are discontinuous at every heterojunction, the numerical fluxes at grid points need to be chosen with care.

Recombination

The recombination rate $R(n(z), p(z); z)$ is a nonlinear function of $n(z)$ and $p(z)$. Straightforward application of HDG would result in the need to repeatedly compute the integral

$$I_{\gamma,q}^h = \int_{K_\gamma^h} R(n_h(z), p_h(z); z)\psi_{\gamma,q}^h(z)\,dz \tag{3.65}$$

for all finite-element shape functions $\psi_{\gamma,q}^h \in \mathbb{V}_h$ during successive steps of the Newton–Raphson method used to solve the discretized nonlinear system of equations. To avoid this computational expense and also simplify calculations, the following interpolatory quadrature scheme is applied element by element [18, 19]. On an element K_γ^h, the Gauss–Lobbato quadrature points $\{z_{\gamma,p}^{GL,h}\}_{p=0}^{P_{\deg}}$ mentioned earlier together with the corresponding Gauss–Lobatto quadrature weights $\{\varrho_{\gamma,p}^{GL,h}\}_{p=0}^{P_{\deg}}$ are used. Then, using the Gauss–Lobatto quadrature, we have

$$I_{\gamma,q}^h \approx \sum_{p=0}^{P_{\deg}} R\left(n_h(z_{\gamma,p}^{GL,h}), p_h(z_{\gamma,p}^{GL,h}); z_{\gamma,p}^{GL,h}\right)\psi_{\gamma,q}^h(z_{\gamma,p}^{GL,h})\varrho_{\gamma,p}^{GL,h} \tag{3.66}$$

for $q \in [0, P_{\deg}]$. The summation simplifies to deliver

$$I_{\gamma,q}^h \approx \sum_{p=0}^{P_{\deg}} R(n_{\gamma,q}^h, p_{\gamma,q}^h; z_{\gamma,q}^{GL,h})\varrho_{\gamma,q}^{GL,h}, \tag{3.67}$$

where $\{n_{\gamma,q}^h\}_{q=0}^{P_{\deg}}$ and $\{p_{\gamma,q}^h\}_{q=0}^{P_{\deg}}$ are the coefficients of n_h and p_h, respectively, on K_γ^h.

Upwinding

To improve the stability of the HDG method, the penalty parameters, akin to those in Eqs. (3.62), appearing in the discretization of $\hat{J}_p$ and $\hat{J}_n$ must be chosen carefully [6]. For $\hat{J}_n$, depending on the sign of $(d/dz)(\phi + \phi_n)$ at the two vertices of each element K_γ^h, the penalty parameters must be chosen to bias the numerical fluxes to the left or right—i.e., either $\tau_\gamma^- > \tau_\gamma^+$ or $\tau_\gamma^- < \tau_\gamma^+$, as appropriate. This corresponds to choosing the penalty parameters depending on the drift of holes, and is why we refer to it as *upwinding*. The penalty parameters have to be chosen in the reverse manner for $\hat{J}_p$; i.e., either $\tau_\gamma^- < \tau_\gamma^+$ or $\tau_\gamma^- > \tau_\gamma^+$.

A possible improvement would be to use an exponential weighting of the Sharfetter–Gummel type [11] to gradually bias the choice of the penalty parameters for the numerical fluxes depending on the magnitudes of the terms in the equations that govern the drift of electrons and holes.

Heterojunctions

Both $n(z)$ and $p(z)$ are generally discontinuous at every heterojunction. This complicates slightly the choice of the vertex terms $\hat{n}_h$ and $\hat{p}_h$ representing the vertex values of $n(z)$ and $p(z)$, respectively. Hence, both $\hat{n}_h$ and $\hat{p}_h$ have to have two values each at every vertex (i.e., the limits from the left and right), and these have to be related using Eqs. (3.47) or (3.48), as appropriate.

Homotopy Method

The result of using the HDG method to discretize Eqs. (3.32a)–(3.32c) is a set of $6N_z(P_{deg} + 1) + 3(N_z + 1)$ equations in the same number of unknown coefficients. The full discretized system is solved using the Newton–Raphson method [20]. One further issue arises: the Newton–Raphson method needs a good initial guess, or it may not converge. In order to obtain a converged solution of the discrete problem, a homotopy method is recommended.

A damping parameter $\delta_o \in (0, 1)$ is chosen to begin the computation of the solution of the electrical submodel. The term $N_f + N_d$ in Eq. (3.61a) and the bandgap variation from its spatially averaged value in Ω_{sc} are multiplied by δ_o. In addition, the term $G(z) - R(n, p; z)$ on the right sides of Eqs. (3.32a) and (3.32b) is replaced by $\delta_o [G(z) - \delta_o R(n, p; z)]$. Then, the Newton–Raphson method is used on the damped system. If convergence is obtained, the value of δ_o is increased by a multiplicative factor toward unity and the Newton–Raphson method is again tried using the output of the previous step as a starting guess. If either convergence does not occur or negative charge-carrier densities are computed, then δ_o is decreased and the Newton–Raphson method is again used with the last successful initial guess. The HDG solution is obtained when $\delta_o = 1$.

The homotopy process is usually successful in getting a convergent HDG solution, but the process can be very time consuming as different values of the damping parameter δ_o are investigated. An understanding of the best choice of which terms to damp in the system might improve the application of the Newton–Raphson method to the HDG method considerably.

By utilizing $J_{n,h}(z)$ and $J_{p,h}(z)$, the finite-element approximations to J_n and J_p, respectively, all solar-cell functions (such as $n(z)$, $p(z)$, $J_n(z)$, etc.) are computed and extracted for post-processing. Repeated solves using different values of V_{ext} also allow the calculation of J_{dev}, J_{sc}, V_{oc}, η and FF. Numerical tests with $P_{deg} = 5$ suggest fourth-order convergence of these parameters, so the convergence rate is suboptimal [6]. Analysis of the convergence of the HDG method for a related time-dependent model drift-diffusion equation [21] proves that a different choice of HDG spaces gives optimal convergence; hence, there is scope for computational improvement with the HDG framework.

3.5   DIFFERENTIAL EVOLUTION ALGORITHM

The design goal is to maximize the power-conversion efficiency η defined by Eq. (3.51). Let the column vector $\tilde{x}$ comprise the $\tilde{N}$ parameters that control the performance of the solar cell: the thicknesses of certain layers in the device, the period of the grating region, the spatial variation of the bandgap in the photon-absorbing layer(s), and so on. In Chapters 4–6, $\tilde{N}$ rarely exceeds ten. Obviously, for any given choice of the control vector $\tilde{x}$ we can first implement the optical submodel to determine $G(\tilde{\mathbf{x}})$ and then the electrical submodel to determine $\eta(\tilde{\mathbf{x}})$, so the final requirement for solar-cell design is a suitable optimization algorithm.

A gradient-free approach incorporating a stochastic search strategy for the optimal control vector is convenient for that purpose. The use of a gradient-free approach allows changing the input parameters in $\tilde{\mathbf{x}}$ without the extensive programming needed to implement each case of a gradient-based optimization code. while the stochastic aspect of the search strategy helps to avoid local maxima. The differential evolution algorithm (DEA) [22] is attractive as it is easily parallelized, uses computer memory efficiently, and is computationally tractable for the dimensionality (i.e., $\tilde{N}$) and the extent of the parameter space in which the design is optimized [23]. In addition, the DEA allows the imposition of upper and lower bound (box) constraints on each of the $\tilde{N}$ control parameters in $\tilde{\mathbf{x}}$, which is important to keep the optimal solution in a physically realizable regime. The main drawback of DEA is the need for many evaluations of $G(\tilde{\mathbf{x}})$ and $\eta(\tilde{\mathbf{x}})$ for different choices of $\tilde{\mathbf{x}}$ as the search probes the parameter space. Even though the control parameters are constrained in the DEA to discrete values within the box constraints so that there is a finite number of possible choices for $\tilde{\mathbf{x}}$, the set of all possible choices is so large that it cannot be investigated exhaustively.

The objective function $\eta : \mathbb{S} \to \mathbb{R}$, where $\mathbb{S} \subset \mathbb{R}^{\tilde{N}}$ is the set of all possible choices of the $\tilde{N}$ control parameters within the box constraints. The DEA starts by selecting N_{P} different control vectors $\tilde{\mathbf{x}}$ to form an initial population $\mathfrak{P}_0 \subset \mathbb{S}$; Storn and Price [22] suggest $N_{\mathrm{P}} \in [5\tilde{N}, 10\tilde{N}]$. The objective function η is evaluated for each of the control vectors making up the population $\mathfrak{P}_0$, with the results used by the DEA in a mutation-recombination-selection process to build a new population $\mathfrak{P}_1 \subset \mathbb{S}$. The objective function is then evaluated for each control vector in this new population to develop a new population $\mathfrak{P}_2 \subset \mathbb{S}$ in the second mutation-recombination-selection step. This process continues iteratively until the objective function appears to stabilize at a maximum.

A Matlab® version of DEA is very convenient [25]. Given a crossover fraction $\tilde{C}$ and stepsize $\tilde{F}$, the steps of the basic algorithm are as follows [23]:

1: Randomly initialize a set of N_{P} control vectors: $\mathfrak{P}_0 = \{\tilde{\mathbf{x}}^{(m)}\}_{m=1}^{N_{\mathrm{P}}} \subset \mathbb{S}$ and set $\tilde{q} = 0$.

2: **while** stopping criterion is not satisfied **do**

3:    Set $\mathfrak{P}_{\tilde{q}+1} = \mathfrak{P}_{\tilde{q}}$

4:    **for** each $\tilde{\mathbf{x}}^{(\nu)} \in \mathfrak{P}_{\tilde{q}+1}$, **do**

5:       Randomly choose three different $\tilde{\mathbf{x}}^{(m_1)}, \tilde{\mathbf{x}}^{(m_2)}, \tilde{\mathbf{x}}^{(m_3)} \in \mathfrak{P}_{\tilde{q}+1}$

6:       Choose a random index $j \in \{1, ..., \tilde{N}\}$

7:       **for** all $\ell \in \{1, ..., \tilde{N}\}$ **do**

8:          Pick $\tilde{r}_\ell$ from uniformly distributed random numbers in $[0, 1]$.

9:          **if** $\tilde{r}_\ell < \tilde{C}$ or $\ell = j$ **then**

10:             $\tilde{\mathbf{w}}_\ell \leftarrow \tilde{\mathbf{x}}_\ell^{(m_1)} + \tilde{F}\left(\tilde{\mathbf{x}}_\ell^{(m_2)} - \tilde{\mathbf{x}}_\ell^{(m_3)}\right)$

11:          **else**

12:             $\tilde{\mathbf{w}}_\ell \leftarrow \tilde{\mathbf{x}}_\ell^{(\nu)}$

13:      **end if**
14:    **end for**
15:    **if** $\eta(\tilde{\mathbf{w}}) > \eta(\tilde{\mathbf{x}}^{(\nu)})$ **then**
16:      $\tilde{\mathbf{w}}$ replaces $\tilde{\mathbf{x}}^{(\nu)}$ in $\mathfrak{P}_{\tilde{q}+1}$
17:    **end if**
18: **end for**
19: $\tilde{q} \leftarrow \tilde{q} + 1$
20: The current $\tilde{\mathbf{x}}^{\mathrm{opt}} \in \mathfrak{P}_{\tilde{q}}$ is a vector for which $\eta(\tilde{\mathbf{x}}^{\mathrm{opt}}) \geq \eta(\tilde{\mathbf{x}})$ for all $\tilde{\mathbf{x}} \in \mathfrak{P}_{\tilde{q}}$
21: **end while**

As the DEA perturbs the current population members with the scaled differences of randomly selected and distinct population members, no separate probability distribution is required to generate the new population [24]. The stopping criterion can include a sufficiently small change to the cost function as well as a maximum number of steps. Following DEA documentation, the crossover fraction $\tilde{C} = 0.6$ and the stepsize $\tilde{F}$ was drawn from uniformly randomly distributed numbers in $[0.5, 1]$ [6] for all numerical results presented in Chapters 4–6.

3.6   BIBLIOGRAPHY

[1] C. Kittel, *Introduction to Solid State Physics*, 8th ed. (Wiley, Hoboken, NJ, 2005). 33

[2] S. M. Sze, *Semiconductor Devices: Physics and Technology*, 2nd ed. (Wiley, New York, NY, 2002). 33, 34, 37

[3] J. Nelson, *The Physics of Solar Cells* (Imperial College Press, London, UK, 2003). 33, 34, 35, 36, 37, 39, 40, 41, 42

[4] S. J. Fonash, *Solar Cell Device Physics*, 2nd ed. (Academic Press, Burlington, MA, 2010). 33, 39, 40, 41

[5] F. Brezzi, L. D. Marini, S. Micheletti, P. Pietra, R. Sacco, and S. Wang, Discretization of semiconductor device problems (I), *Handbook of Numerical Analysis: Numerical Methods for Electrodynamic Problems Vol. 13*, W. H. A. Schilders and E. J. W. ter Maten, Eds., pages 317–441, Elsevier, Amsterdam, The Netherlands (2005). 39

[6] T. H. Anderson, B. J. Civiletti, P. B. Monk, and A. Lakhtakia, Coupled optoelectronic simulation and optimization of thin-film photovoltaic solar cells, *Journal of Computational Physics*, 407:109242 (2020). 40, 41, 42, 43, 45, 47, 48, 49, 51

T. H. Anderson, B. J. Civiletti, P. B. Monk, and A. Lakhtakia, Coupled optoelectronic simulation and optimization of thin-film photovoltaic solar cells, *Journal of Computational Physics*, 418:109561 (2020) (corrigendum).

[7] D. H. Foster, T. Costa, M. Peszynska, and G. Schneider, Multiscale modeling of solar cells with interface phenomena, *Journal of Coupled Systems and Multiscale Dynamics*, 1:179–204 (2013). 42

[8] K. Yang, J. R. East, and G. I. Haddad, Numerical modeling of abrupt heterojunctions using a thermionic-field emission boundary condition, *Solid-State Electronics*, 36:321–330 (1993). 42

[9] D. L. Scharfetter and H. K. Gummel, Large-signal analysis of a silicon read diode oscillator, *IEEE Transactions on Electron Devices*, 16:64–77 (1969). 44

[10] F. Brezzi, L. D. Marini, and P. Pietra, Two-dimensional exponential fitting and applications to drift-diffusion models, *SIAM Journal on Numerical Analysis*, 26:1342–1355 (1989). 44

[11] P. A. Farrell and E. C. Gartland Jr., On the Scharfetter–Gummel discretization for drift-diffusion continuity equations, *Computational Methods for Boundary and Interior Layers in Several Dimensions*, J. J. H. Miller, Ed., pages 51–79, Boole Press, Dublin, Ireland (1991). 44, 48

[12] D. Brinkman, K. Fellner, P. Markowich, and M.-T. Wolfram, A drift-diffusion-reaction model for excitonic photovoltaic bilayers: Asymptotic analysis and a 2-D HDG finite-element scheme, *Mathematical Models and Methods in Applied Sciences*, 23:839–872 (2013). 44

[13] B. Cockburn, J. Gopalakrishnan, and R. Lazarov, Unified hybridization of discontinuous Galerkin, mixed, and continuous Galerkin methods for second order elliptic problems, *SIAM Journal on Numerical Analysis*, 47:1319–1365 (2009). 44, 45

[14] G. Fu, W. Qiu, and W. Zhang, An analysis of HDG methods for convection-dominated diffusion problems, *ESAIM: Mathematical Modelling and Numerical Analysis*, 49:225–256 (2015). 44

[15] C. Lehrenfeld, Hybrid discontinuous Galerkin methods for solving incompressible flow problems, Diplomingenieur Dissertation (Rheinisch-Westfäälischen Technischen Hochschule, Aachen, Germany, 2010). 44

[16] Y. Chen, P. Kivisaari, M.-E. Pistol, and N. Anttu, Optimization of the short-circuit current in an InP nanowire array solar cell through opto-electronic modeling, *Nanotechnology*, 27:435404 (2016).

[17] J. S. Hesthaven, *Numerical Methods for Conservation Laws: From Analysis to Algorithm* (SIAM, Philadelphia, PA, 2018). 44

[18] J. Douglas Jr. and T. Dupont, The effect of interpolating the coefficients in nonlinear parabolic Galerkin procedures, *Mathematics of Computation*, 29:360–389 (1975). 48

[19] B. Cockburn, J. R. Singler, and Y. Zhang, Interpolatory HDG method for parabolic semilinear PDEs, *Journal of Scientific Computing*, 79:1777–1800 (2019). 48

[20] Y. Jaluria, *Computer Methods for Engineering* (Taylor and Francis, Washington, DC, 1996). 49

[21] G. Chen, P. Monk, and Y. Zhang, An HDG Method for the time-dependent drift-diffusion model of semiconductor devices, *Journal of Scientific Computing*, 80:420–443 (2019). 49

[22] R. Storn and K. Price, Differential evolution—a simple and efficient heuristic for global optimization over continuous spaces, *Journal of Global Optimization*, 11:341–359 (1997). 50

[23] A. Slowik and H. Kwasnicka, Evolutionary algorithms and their applications to engineering problems, *Neural Computing and Applications*, 32:12363–12379 (2020). 50

[24] S. Das and P. N. Suganthan, Differential evolution: A survey of the state-of-the-art, *IEEE Transactions on Evolutionary Computation*, 15:4–31 (2011). 51

[25] M. Buehren, *Differential Evolution* (accessed June 29, 2021). 50

CHAPTER 4

Homogeneous Photon-Absorbing Layer

4.1 CIGS SOLAR CELLS

$CuIn_{1-\xi}Ga_\xi Se_2$ is a quaternary I-III-VI semiconductor formed as a solid solution of the ternary compounds $CuInSe_2$ and $CuGaSe_2$. CIGS has a chalcopyrite structure in which indium and gallium atoms occupy the same atomic sites. The CIGS bandgap varies as a function of the parameter ξ, with Ga and In being present in the ratio $\xi : (1 - \xi)$. The most efficient thin-film CIGS solar cells are obtained when the CIGS bandgap lies between 1.15 eV and 1.25 eV [1]. The optoelectronic properties of CIGS strongly depend upon the deposition technique. The most widely used method for the deposition of the CIGS absorber layer is the physical vapor deposition of the metallic elements (copper, indium, and gallium) and the sole non-metallic element (selenium) at high temperatures [2].

The maximum power-conversion efficiency η of thin-film CIGS solar cells is 22.6% [2, 3], the CIGS photon-absorbing layer being the major contributor toward the electron-hole-pair generation rate. The maximum efficiency of CIGS solar cells compares well with the 26.7% maximum efficiency of single-junction c-Si solar cells [3]. However, the scarcity of indium is a significant obstacle for large-scale and low-cost production of thin-film CIGS solar cells [4]. If the thickness of the CIGS layer could be reduced without significantly reducing the efficiency, this obstacle could be overcome. Thinning of the photon-absorbing layer will also reduce the electron-hole-pair recombination rate R and improve the open-circuit voltage V_{oc} [5]. Furthermore, thickness reduction will enhance manufacturing throughput. However, naïvely reducing the CIGS layer thickness below 1000 nm can lower η for two reasons. First, the lower absorption of solar photons would reduce both the optical short-circuit current density J_{sc}^{opt} and V_{oc} [6]. Second, the increase of R near the back-contact [7, 8] would reduce the device current density J_{dev}.

The CIGS layer must therefore be designed carefully to improve photon absorption therein and limit the back-contact recombination for higher efficiency. Furthermore, as a periodic nanostructured array of silica (SiO_2) nano-particles coated on the back-contact was found experimentally to improve the efficiency of ultrathin CIGS solar cells [9, 10], the effect of a periodically corrugated back reflector may also be considered [11], as discussed in Section 1.5.2.

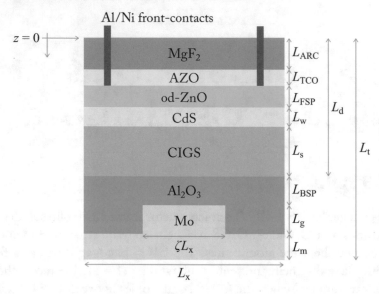

Figure 4.1: Schematic of the reference unit cell of the CIGS solar cell with a homogeneous photon-absorbing layer and a periodically corrugated back reflector. The corrugation height $L_g = 0$ for the standard solar cell.

4.1.1   STRUCTURE

The standard CIGS solar cell is shown schematically in Fig. 4.1 [2], with magnesium fluoride (MgF$_2$) as an anti-reflection coating [12], aluminum-doped zinc oxide (AZO) as the front-contact to collect electrons [13], oxygen-deficient zinc oxide (od-ZnO) [14] and cadmium sulfide (CdS) [15] together forming a bilayer buffer that serves as an n-type semiconductor, a p-type CIGS photon-absorbing layer, an aluminum oxide (Al$_2$O$_3$) [16] layer for back-surface passivation, and molybdenum (Mo) as the back-contact to collect holes. The commonly used substrate for the deposition of different layers is a soda-lime glass which is attractive for its low cost and high stability at higher temperatures [17].

Defects form on the CdS/CIGS interface. Guillemoles et al. [18] found that the n-type indium-rich surface layer formed at the CdS/CIGS interface is beneficial for CIGS solar cells due to the shift of the electrical junction away from the high-recombination interface of the CdS and CIGS layers. In contrast, high interfacial-defect density might lead to deterioration of the solar-cell performance [19, 20]. However, the coupled optoelectronic model developed in Chapters 2 and 3 has shown that a 10-nm-thick surface-defect layer between CdS and CIGS reduces η by an insignificant margin [11], which result is in accord with an experimental study on high-efficiency CIGS solar cells [21]. Hence, the surface defects can be ignored for modeling and design.

A thin Al_2O_3 back-surface passivation layer between the CIGS layer and the Mo back reflector has been found to reduce the recombination rate in the vicinity of the CIGS/Mo interface by chemical and field-effect passivation. It increases both J_{sc} and V_{oc} compared to the unpassivated CIGS solar cell, as has been found both experimentally [22–25] and by simulation [11].

Although excellent for the back electrode, Mo is not the best choice for a back reflector. Tungsten and tantalum improve optical reflection and enhance efficiency compared to Mo, but remain limited to solar cells of small sizes [26]. Despite higher reflectance, titanium, vanadium, chromium, and manganese deteriorate the solar-cell performance due to their reactivity with selenium during the CIGS growth process [8]. Gold and silver diffuse into the CIGS layer at high temperatures during deposition [8]. Furthermore, silver is less stable than Mo at temperatures exceeding 550°C required for the fabrication of CIGS solar cells. Hence, Mo continues to be used in CIGS solar cells.

4.1.2   GEOMETRIC DESCRIPTION

The solar cell occupies the region $\mathcal{X} : \{(x, y, z)| -\infty < x < \infty, -\infty < y < \infty, 0 < z < L_t\}$, with the half spaces $z < 0$ and $z > L_t$ occupied by air. The reference unit cell of this structure is identified as $\mathcal{R} : \{(x, y, z)| -L_x/2 < x < L_x/2, -\infty < y < \infty, 0 < z < L_t\}$, with the back reflector being periodically corrugated with period L_x along the x axis.

The typical thicknesses of the top four layers are as follows: $L_{ARC} = 110$ nm (MgF_2), $L_{TCO} = 100$ nm (AZO), $L_{FSP} = 80$ nm (od-ZnO), and $L_w = 70$ nm (CdS). The 80-nm-thick layer of od-ZnO serves as a buffer layer to increase the open-circuit voltage [27]. Oxygen deficiency during its deposition makes ZnO an n-type semiconductor [14]. CdS is also an n-type semiconductor, whereas the CIGS layer of thickness $L_s \leq 2200$ nm is a p-type semiconductor. The layer of thickness $L_{BSP} = 50$ nm is made of Al_2O_3 [16] to protect the electrical characteristics of the CIGS layer and it also functions as a passivation layer. The layer of thickness $L_m = 500$ nm is made of Mo, its thickness being well beyond the electromagnetic penetration depth [28] in the optical regime. The region $L_d + L_{BSP} < z < L_d + L_{BSP} + L_g$ is a rectangular Mo/Al_2O_3 grating with period L_x along the x axis. The corrugation height $L_g = 0$ for the standard solar cell.

4.1.3   OPTICAL DESCRIPTION

The permittivity in the grating region of $\mathcal{R}$ is given by

$$\varepsilon_g(x, z, \lambda_0) = \begin{cases} \varepsilon_m(\lambda_0), & |x| < \zeta L_x/2, \\ \varepsilon_d(\lambda_0), & |x| > \zeta L_x/2, \end{cases}$$
$$z \in (L_d + L_{BSP}, L_d + L_{BSP} + L_g), \qquad (4.1)$$

where $\zeta \in [0, 1]$ is the duty cycle, $\varepsilon_m(\lambda_0)$ is the permittivity of molybdenum [29] and $\varepsilon_d(\lambda_0)$ is the permittivity of Al_2O_3 [16].

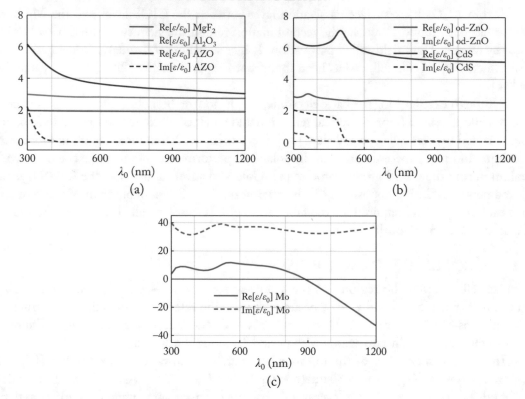

Figure 4.2: Real and imaginary parts of the relative permittivity $\varepsilon/\varepsilon_0$ of (a) MgF_2, Al_2O_3, and AZO; (b) od-ZnO and CdS; and (c) Mo as functions of λ_0 in the optical regime [11]. The imaginary part of the relative permittivity is negligibly small for both MgF_2 and Al_2O_3.

Spectrums of the real and imaginary parts of the relative permittivity $\varepsilon(\lambda_0)/\varepsilon_0$ of MgF_2 [12], AZO [13], od-ZnO [30], CdS [15], Al_2O_3 [16], and Mo [29] used for calculations are displayed in Fig. 4.2. Spectrums of the real and imaginary parts of the relative permittivity $\varepsilon(\lambda_0)/\varepsilon_0$ of CIGS [31, 32] as functions of ξ are displayed in Fig. 4.3.

4.1.4   ELECTRICAL DESCRIPTION

The region $L_{ARC} + L_{TCO} < z < L_d + L_{BSP}$ (Fig. 4.1) containing the od-ZnO, CdS, CIGS, and Al_2O_3 layers must be considered for electrical analysis, because these four layers lie between the front-contact and the back-contact layers. Whereas Al_2O_3 is an insulator, both od-ZnO and CdS contribute to charge-carrier generation. The useful solar spectrum is typically taken to span free-space wavelengths in excess of $\lambda_{0,min} = 300$ nm. Since od-ZnO has a bandgap of 3.3 eV, it can absorb solar photons with energies corresponding to $\lambda_0 \in [300, 376]$ nm. Likewise,

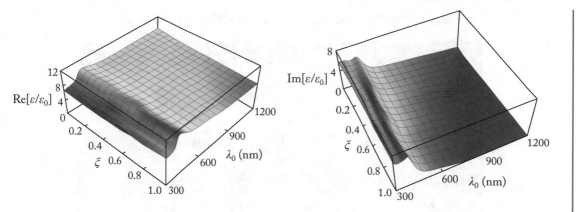

Figure 4.3: Real and imaginary parts of the relative permittivity $\varepsilon/\varepsilon_0$ of CIGS as functions of λ_0 and ξ [11].

as CdS has a bandgap of 2.4 eV, it can absorb solar photons with energies corresponding to $\lambda_0 \in$ [300, 517] nm. Hence, the generation of electron-hole pairs in the od-ZnO and CdS layers must be accounted for, not to mention the recombination of electron-hole pairs in both layers. Values of the electrical parameters of od-ZnO, CdS, and CIGS [19] are provided in Table 4.1, with the exception of the Auger-recombination parameters C_n and C_p as their values are unavailable.

4.1.5 COUPLED OPTOELECTRONIC MODELING AND OPTIMIZATION

The coupled optoelectronic model of Chapters 2 and 3 [33] has been validated for CIGS solar cells by comparison with extant experimental results for the conventional MgF$_2$/AZO/od-ZnO/CdS/CIGS/Mo solar cell containing a 2200-nm-thick homogeneous CIGS layer and a flat back reflector [34]. Values of J_{sc}, V_{oc}, FF, and η obtained from the model for $\xi = 0$ (E$_g$ = 0.947 eV), $\xi = 0.25$ (E$_g$ = 1.12 eV), and $\xi = 1$ (E$_g$ = 1.626 eV) are provided in Table 4.2 [11]. The model predictions are in reasonable agreement with the experimental data [2, 34], the differences being very likely due to variance between the optical and electrical properties used in the model from those realized in practice.

Flat Back Reflector

The coupled optoelectronic model and the differential evolution algorithm have been used to maximize η as a function of the bandgap energy E$_g$ $\in$ [0.947, 1.626] eV and the thickness L_s $\in$ [100, 2200] nm of the CIGS layer when the back reflector is flat (i.e., L_g = 0) [11]. Values of J_{sc}, V_{oc}, FF, and η [35, 36] corresponding to the optimal design for $L_s \in$ {100, 200, 300, 400, 500, 600, 900, 1200, 2200} nm are shown in Table 4.3. Depending on L_s, the optimal homogeneous bandgap varies, with E$_g$ $\in$ [1.24, 1.28] eV. The optimal efficiency in-

Table 4.1: Electrical parameters of od-ZnO, CdS, and CIGS for $\xi \in [0, 1]$

Parameter (unit)	od-ZnO [19]	CdS [19]	CIGS [19]
E_g (eV)	3.3	2.4	$0.947 + 0.679\xi$
χ (eV)	4.4	4.2	$4.5 - 0.6\xi$
N_c (cm^{-3})	3×10^{18}	1.3×10^{18}	6.8×10^{17}
N_v (cm^{-3})	1.7×10^{19}	9.1×10^{19}	1.5×10^{19}
N_D (cm^{-3})	1×10^{17}	5×10^{17}	
N_A (cm^{-3})			2×10^{16}
μ_n (cm^2 V^{-1} s^{-1})	100	72	100
μ_p (cm^2 V^{-1} s^{-1})	31	20	13
ε_{dc}	9	5.4	13.6
N_f (cm^{-3})	10^{16}	5×10^{17}	$10^{13}(1 + 10^3\xi)$
E_T	$0.5E_g$	$0.5E_g$	$0.5E_g$
σ_n (cm^2)	5×10^{-13}	5×10^{-13}	5×10^{-13}
σ_p (cm^2)	10^{-15}	10^{-15}	10^{-15}
R_B (cm^3 s^{-1})	10^{-10}	10^{-10}	10^{-10}
$v_{th} = v_{th,n} = v_{th,p}$ (cm s^{-1})	10^7	10^7	10^7

creases with L_s. An efficiency of 13.79% is predicted with a 600-nm-thick CIGS layer. The highest efficiency predicted is 18.93% for a solar cell with a 2200-nm-thick CIGS layer with an optimal bandgap of $E_g = 1.24$ eV.

Periodically Corrugated Back Reflector

Optoelectronic optimization for solar cells containing a homogeneous CIGS layer but with a periodically corrugated back reflector instead of a flat one has also been performed [11], the four-dimensional (i.e., $\tilde{N} = 4$) parameter space for optimizing η being as follows: $E_g \in [0.947, 1.626]$ eV, $L_g \in [1, 550]$ nm, $\zeta \in (0, 1)$, and $L_x \in [100, 1000]$ nm. The results of this optimization exercise for fixed L_s are provided in Table 4.4.

On comparing Tables 4.3 and 4.4, it is clear that the periodic corrugation of the back reflector improves the efficiency by no more than 2% (at $L_s = 200$ nm). This indicates the slight benefit of exciting both surface-plasmon-polariton waves [37–40] and waveguide modes [41, 42] by taking advantage of the grating corrugations [43], when the CIGS layer is ultrathin. However, that benefit vanishes for thicker CIGS layers.

Table 4.2: Comparison of J_{sc}, V_{oc}, FF, and η predicted by the coupled optoelectronic model [11] for conventional CIGS solar cells with a 2200-nm-thick CIGS photon-absorbing layer and a flat back reflector with their experimental counterparts [2, 34]

ξ	E_g (eV)		J_{sc} (mA cm^{-2})	V_{oc} (mV)	FF (%)	η (%)
0	0.950	Model [11]	38.63	497	78	15.05
		Experiment [34]	40.58	491	66	14.50
		Experiment [34]	41.10	491	75	15.00
0.25	1.120	Model [11]	34.41	648	81	18.12
		Experiment [34]	35.22	692	79	19.50
		Experiment [2]	37.80	741	81	22.60
1	1.626	Model [11]	14.86	911	73	9.92
		Experiment [34]	14.88	823	71	9.53
		Experiment [34]	18.61	905	75	10.20

Table 4.3: Model-predicted parameters of the optimal CIGS solar cell with a specified value of $L_s \in [100, 2200]$ nm, when the CIGS layer is homogeneous and the back reflector is flat ($L_g = 0$) [11]

L_s (nm)	E_g (eV)	J_{sc} (mA cm^{-2})	V_{oc} (mV)	FF (%)	η (%)
100	1.28	14.89	624	78	7.25
200	1.26	19.50	660	76	9.76
300	1.25	22.56	681	76	11.59
400	1.27	22.65	711	77	12.49
500	1.25	23.71	704	78	13.15
600	1.24	24.66	704	79	13.79
900	1.25	25.68	725	80	15.08
1200	1.28	25.72	756	81	15.90
2200	1.24	31.11	742	82	18.93

Table 4.4: Model-predicted parameters of the optimal CIGS solar cell with a specified value of $L_s \in [100, 2200]$ nm, when the CIGS layer is homogeneous and the back reflector is periodically corrugated [11]

L_s (nm)	E_g (eV)	L_g (nm)	ζ	L_x (nm)	J_{sc} (mA cm^{-2})	V_{oc} (mV)	FF (%)	η (%)
100	1.28	97	0.50	500	14.89	624	78	7.25
200	1.26	101	0.50	510	19.53	661	76	9.91
300	1.25	101	0.50	510	22.56	681	76	11.59
400	1.27	101	0.49	510	22.79	711	77	12.58
500	1.25	106	0.48	510	23.78	705	78	13.19
600	1.24	105	0.48	510	24.69	704	79	13.81
900	1.25	101	0.49	502	25.71	725	80	15.09
1200	1.28	101	0.49	502	25.72	759	81	15.90
2200	1.24	101	0.49	502	31.11	742	82	18.93

4.2   CZTSSE SOLAR CELLS

The scarcity of indium in CIGS solar cells and of tellurium in CdTe solar cells [4], as well as the toxicity of cadmium in CdTe solar cells, motivated exploration of alternative materials that are abundant on our planet, and that can be extracted, processed, and discarded with a low environmental cost. $Cu_2ZnSn(S_\xi Se_{1-\xi})_4$ is a p-type I_2–II–IV–VI_2 semiconductor that can be made from nontoxic and Earth-abundant materials [44] for use in place of CIGS in a solar cell, as can be seen by comparing Figs. 4.1 and 4.4. The bandgap of CZTSSe can be varied by controlling the compositional parameter $\xi \in [0, 1]$.

The maximum power-conversion efficiency η of thin-film CZTSSe solar cells is only 12.6% [45, 46], which is substantially lower than the 22.6% record efficiency of CIGS solar cells [2, 3]. This can be attributed to the existence of more bandtail states in CZTSSe [47, 48], higher electron-hole-pair recombination inside the CZTSSe layer because of the shorter lifetime of minority carriers (electrons) [49], and the higher recombination at the p-n junction [50].

In contrast to CIGS, CZTSSe has more defects due to the higher number of constituent elements and the formation of secondary phases such as SnS_2 and Sn_2S_3 before or during fabrication. For example, Zn cations can occupy Cu sites and vice versa due to their similar ionic radii [47]. These defects and secondary phases cause more bandtail states and thereby lower the bandgap energy E_g [48]. Also, the short lifetime of minority carriers reduces their diffusion length, thereby limiting the collection of minority carriers deep in the CZTSSe absorber layer [49, 51]. For example, the diffusion length of electrons is less than 1000 nm when the bandgap energy E_g of CZTSSe is 1.15 eV (for $\xi \approx 0.41$), which means that a solar cell with a CZTSSe layer of thickness $L_s > 1000$ nm [51] will have a high series resistance [52, 53]. Higher

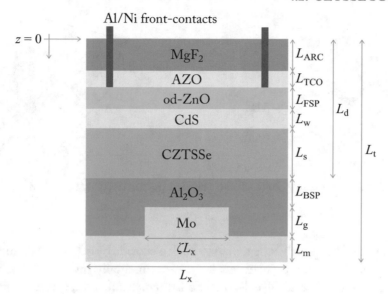

Figure 4.4: Schematic of the reference unit cell of the CZTSSe solar cell with a homogeneous photon-absorbing layer and a periodically corrugated Mo back reflector. The corrugation height $L_g = 0$ for the standard solar cell.

recombination at the *p-n* junction reduces the surface current density. All of these factors lower V_{oc}, thereby lowering η [54, 55].

4.2.1   STRUCTURE

The structure of CZTSSe solar cells is almost the same as of CIGS solar cells, i.e., MgF_2/AZO/od-ZnO/CdS/CZTSSe/Al_2O_3/Mo. The descriptions of all layers in this structure are the same as for the CIGS solar cell discussed in Section 4.1.1 except that the CIGS layer is replaced with the CZTSSe layer.

One of the reasons for the low open-circuit voltage of CZTSSe solar cells is a higher recombination rate at the CdS/CZTSSe interface [45, 53, 56]. The incorporation of a 10-nm-thick surface-defect layer [20] to account for the higher recombination rate was found to barely affect the model-predicted efficiency [57]. Hence, the surface defects can be ignored for modeling and design.

The deposition of CZTSSe on Mo creates an interfacial $Mo(S_\xi Se_{1-\xi})_2$ layer, which enhances the electron-hole recombination rate in the vicinity of the Mo back electrode, thereby depressing η [58]. The insertion of a thin Al_2O_3 layer between CZTSSe and Mo improves η by (i) suppressing the decomposition reactions which cause high device resistance and recombination rate and (ii) preventing the formation of the $Mo(S_\xi Se_{1-\xi})_2$ layer [24, 58].

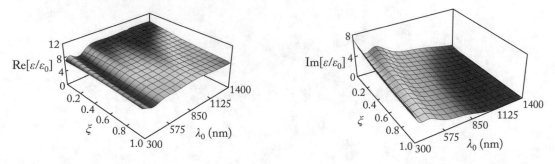

Figure 4.5: Real and imaginary parts of the relative permittivity $\varepsilon/\varepsilon_0$ of CZTSSe as functions of λ_0 and ξ [57].

Replacement of the poorly reflective Mo back electrode by other metals such as tungsten, gold, palladium, platinum, and nickel has been experimentally investigated [59]. Palladium, platinum, and nickel deteriorate the solar-cell performance due to their reactions with selenium during the CZTSSe fabrication process. Tungsten and gold are better reflectors, but Mo has superior charge-collection ability. Hence, Mo remains the the best back-electrode material for power-conversion efficiency.

4.2.2   GEOMETRIC DESCRIPTION

The CZTSSe solar cell occupies the region $\mathcal{X} : \{(x, y, z)| -\infty < x < \infty, -\infty < y < \infty,$ $0 < z < L_t\}$, the half spaces $z < 0$ and $z > L_t$ being occupied by air. The reference unit cell of this structure is shown in Fig. 4.4.

The typical thicknesses of the top four layers are as follows: $L_{\mathrm{ARC}} = 110$ nm (MgF$_2$), $L_{\mathrm{TCO}} = 100$ nm (AZO), $L_{\mathrm{FSP}} = 100$ nm (od-ZnO), and $L_{\mathrm{w}} = 50$ nm (CdS). Both od-ZnO and CdS are n-type semiconductors [14, 15], whereas CZTSSe is a p-type semiconductor. The Al$_2$O$_3$ layer of thickness $L_{\mathrm{BSP}} = 20$ nm plays multiple roles [24, 58]. The layer of thickness $L_{\mathrm{m}} = 500$ nm is made of Mo, its thickness being well beyond the electromagnetic penetration depth [28]. The region $L_d + L_{\mathrm{BSP}} < z < L_d + L_{\mathrm{BSP}} + L_g$ is a rectangular Mo/Al$_2$O$_3$ grating with duty cycle $\zeta \in (0, 1)$, height L_g, and period L_x along the x axis. The grating is absent for $\zeta \in \{0, 1\}$.

4.2.3   OPTICAL DESCRIPTION

The permittivity in the grating region of $\mathcal{R}$ is given by Eq. (4.1). Spectrums of the real and imaginary parts of the relative permittivity $\varepsilon(\lambda_0)/\varepsilon_0$ of MgF$_2$ [12], AZO [13], od-ZnO [30], CdS [15], Al$_2$O$_3$ [16], and Mo [29] are displayed in Fig. 4.2. Spectrums of the real and imaginary parts of the relative permittivity $\varepsilon/\varepsilon_0$ of CZTSSe [32, 44, 60, 61] as functions of ξ are displayed in Fig. 4.5.

Table 4.5: Electrical properties of od-ZnO, CdS, and CZTSSe for $\xi \in [0, 1]$

Parameter (unit)	od-ZnO [19]	CdS [19]	CZTSSe [44, 54]
E_g (eV)	3.3	2.4	$0.91 + 0.58\xi$ (optical submodel)
			$0.91 + 0.44\xi$ (electrical submodel)[†]
χ (eV)	4.4	4.2	$4.46 - 0.16\xi$
N_c (cm^{-3})	3×10^{18}	1.3×10^{18}	7.8×10^{17}
N_v (cm^{-3})	1.7×10^{19}	9.1×10^{19}	4.5×10^{18}
N_D (cm^{-3})	1×10^{17}	5×10^{17}	
N_A (cm^{-3})			1×10^{16}
μ_n (cm^2 V^{-1} s^{-1})	100	72	40
μ_p (cm^2 V^{-1} s^{-1})	31	20	12.6
ε_{dc}	9	5.4	$14.9 - 1.2\xi$
N_f (cm^{-3})	10^{16}	5×10^{17}	$(1.35 + 98.65\xi) \times 10^{15}$
E_T	$0.5E_g$	$0.5E_g$	$0.5E_g$
σ_n (cm^2)	5×10^{-13}	5×10^{-13}	10^{-14}
σ_p (cm^2)	10^{-15}	10^{-15}	10^{-14}
R_B (cm^3 s^{-1})	10^{-10}	10^{-10}	10^{-10}
$v_{th} = v_{th,n} = v_{th,p}$ (cm s^{-1})	10^7	10^7	10^7

[†] E_g is artificially reduced from its value in the optical submodel by 0.14ξ eV for the electrical model so as to account for bandtail states.

4.2.4 ELECTRICAL DESCRIPTION

For electrical modeling, only the region $L_{ARC} + L_{TCO} < z < L_d + L_{BSP}$ has to be considered. Electron-hole-pair generation occurs only in the CZTSSe, od-ZnO, and CdS layers. The generation rate $G(z)$ contains the effects of (i) the periodic corrugations of the back reflector, (ii) the Al_2O_3 back-surface passivation layer, and (iii) the MgF_2 anti-reflection coating. Table 4.5 provides the values of electrical parameters used for od-ZnO, CdS, and CZTSSe [19, 44, 54]. The effect of bandtail states, which effectively narrow the bandgap, is incorporated [47, 48] by reducing the bandgap of CZTSSe for electrical calculations. Whereas $E_g = 0.91 + 0.58\xi \in [0.91, 1.49]$ eV has to be used for CZTSSe in the optical submodel of the coupled optoelectronic model, $E_g = 0.91 + 0.44\xi \in [0.91, 1.35]$ eV has to be used in the electrical submodel [47, 48]. Values of the Auger-recombination parameters C_n and C_p of od-ZnO, CdS, and CZTSSe are unavailable.

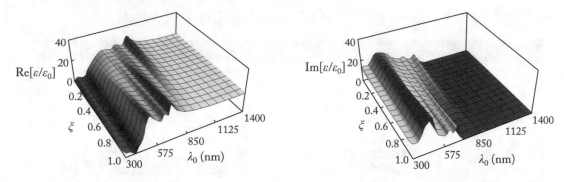

Figure 4.6: Real and imaginary parts of the relative permittivity $\varepsilon/\varepsilon_0$ of $Mo(S_\xi Se_{1-\xi})_2$ as functions of λ_0 and ξ [57].

4.2.5   COUPLED OPTOELECTRONIC MODELING AND OPTIMIZATION

The coupled optoelectronic model of Chapters 2 and 3 [33] has been validated by comparison with experimental results available for the conventional MgF$_2$/AZO/od-ZnO/CdS/CZTSSe/$Mo(S_\xi Se_{1-\xi})_2$/Mo solar cell containing a 2000-nm-thick homogeneous CZTSSe layer and a flat back reflector [45]. In this solar cell, a 200-nm-thick $Mo(S_\xi Se_{1-\xi})_2$ layer with defect density $N_f = 10^{18}$ cm^{-3} replaces the Al$_2$O$_3$ layer in Fig. 4.4, and appropriate modifications were made for the validation. The relative permittivity $\varepsilon(\lambda_0/\varepsilon_0)$ of $Mo(S_\xi Se_{1-\xi})_2$ in the optical regime is displayed in Fig. 4.6. All other relevant electrical parameters of $Mo(S_\xi Se_{1-\xi})_2$ were taken to be the same as that of CZTSSe, except that $E_g = 1.57 + 0.31\xi$ eV [62] was used for $Mo(S_\xi Se_{1-\xi})_2$ in both the optical and electrical submodels in the coupled optoelectronic model.

The values of J_{sc}, V_{oc}, FF, and η obtained from the coupled optoelectronic model for $\xi \in \{0, 0.38, 1\}$ are provided in Table 4.5 along with the corresponding experimental data [45, 48, 53]. According to this table, the model's predictions are in reasonable agreement with the experimental data, the variances being very likely due to differences between the optical and electrical properties inputted to the model from those realized in practice. As interface defects are not explicitly considered in the model, all the experimentally observed features can be adequately accounted for by the bulk properties of CZTSSe, which is also in accord with the empirical model provided by Gokmen et al. [63].

Further insight on the role of the $Mo(S_\xi Se_{1-\xi})_2$ layer comes when its thickness is lowered from 200 nm to 100 nm but the thickness L_s of the CZTSSe layer is increased from 2000 nm to 2100 nm. Ahmad et al. [57] found that the coupled optoelectronic model predicts the efficiency to increase from 11.15% to 11.23%, indicating the minor role of the thickness of the $Mo(S_\xi Se_{1-\xi})_2$ layer. When the 100-nm-thick $Mo(S_\xi Se_{1-\xi})_2$ layer is replaced by a 20-nm-thick

Table 4.6: Comparison of J_{sc}, V_{oc}, FF, and η predicted by the coupled optoelectronic model for conventional MgF$_2$/AZO/od-ZnO/CdS/CZTSSe/Mo(S$_\xi$Se$_{1-\xi}$)$_2$ /Mo solar cells with a 2000-nm-thick CZTSSe photon-absorbing layer, a 200-nm-thick Mo(S$_\xi$Se$_{1-\xi}$)$_2$ layer, and a flat back reflector with experimental counterparts [57]. The Al$_2$O$_3$ layer is absent for these data.

ξ		J_{sc} (mA cm^{-2})	V_{oc} (mV)	FF (%)	η (%)
0	Model [57]	38.31	361.0	65.0	8.96
	Experiment [53]	36.40	412.0	62.0	9.33
0.38	Model [57]	32.42	509.0	69.0	11.15
	Experiment [45]	35.20	513.4	69.8	12.60
1	Model [57]	17.86	606.0	60.7	6.61
	Experiment [48]	16.90	637.0	61.7	6.70

Al$_2$O$_3$ layer, the efficiency changes very slightly. Thus, the growth of an Al$_2$O$_3$ layer for back passivation is highly advisable.

Flat Back Reflector

The coupled optoelectronic model along with the differential evolution algorithm has been used to maximize η as a function of $\xi \in [0, 1]$ (for the optical part) and the thickness $L_s \in [100, 2200]$ nm of the CZTSSe layer when the back reflector is flat (i.e., $L_g = 0$) [57]. Then, $E_g \in [0.91, 1.49]$ eV for the optical submodel and $E_g \in [0.91, 1.35]$ eV for the electrical submodel to account for bandtail states, the difference between the two values of E_g being 0.14ξ eV; see Table 4.5.

With $L_s = 2200$ nm, the maximum efficiency predicted is 11.76% when $\xi = 0.5$. The efficiency generally becomes lower for $\xi > 0.5$ because of the narrowing of the portion of the solar spectrum available for photon absorption [64] due to the blue shift of $\lambda_{0,max}$, and the increased recombination due to increase in N_f caused by the higher value of ξ [44, 53, 54].

In order to compare the performance of the solar cell with optimal L_s, values of ξ, J_{sc}, V_{oc}, FF, and η predicted by the coupled optoelectronic model are presented in Table 4.7 for seven representative values of L_s. The highest efficiency is not exhibited by the thickest photon-absorbing layer. The maximum efficiency increases as L_s increases from 100 nm to 1200 nm, but decreases at a very slow rate with a further increase of L_s. The highest value of η predicted is 11.84% when $L_s = 1200$ nm and $\xi = 0.5172$. The values of J_{sc}, V_{oc}, and FF corresponding to this optimal design are 30.13 mA cm^{-2}, 558 mV, and 70.3%, respectively.

The efficiency increase with L_s for $L_s < 1200$ nm is due to the increase in volume available to absorb photons. The efficiency reduction for $L_s > 1200$ nm is due to reduced charge-carrier collection arising from the diffusion length of minority charge-carriers in CZTSSe being smaller

Table 4.7: Model-predicted parameters of the optimal MgF$_2$/AZO/od-ZnO/CdS/CZTSSe/Al$_2$O$_3$/Mo solar cell with a specified value of $L_s \in [100, 2200]$ nm, when the CZTSSe layer is homogeneous and the back reflector is flat ($L_g = 0$) [57]

Ls (nm)	ξ	J_{sc} (mA cm^{-2})	V_{oc} (mV)	FF (%)	η (%)
100	0.5172	19.23	513	75.2	7.41
200	0.5000	25.19	535	72.0	9.67
300	0.5000	27.27	546	69.6	10.38
400	0.5000	28.07	551	69.6	10.79
600	0.4656	29.31	556	70.0	11.47
1200	0.5172	30.13	558	70.3	11.84
2200	0.5000	30.00	557	70.3	11.76

than L_s [51]. Notably, the optimal bandgap of the CZTSSe layer fluctuates in a small range (i.e., [1.18, 1.21] eV), despite a 22-fold increase of L_s.

Periodically Corrugated Back Reflector

Optoelectronic optimization for solar cells containing a homogeneous CZTSSe layer but with a periodically corrugated back reflector instead of a flat one has also been performed [57], the four-dimensional (i.e., $\tilde{N} = 4$) parameter space for optimizing η being as follows: $\xi \in [0, 1]$, $L_g \in [1, 550]$ nm, $\zeta \in (0, 1)$, and $L_x \in [100, 1000]$ nm. The results of this optimization exercise for fixed L_s are provided in Table 4.7 for seven representative values of L_s. The values of ξ, L_g, ζ, and L_x for the optimal designs are also provided in the same table.

A comparison of Tables 4.7 and 4.8 reveals that periodic corrugation of the Mo back reflector slightly improves η for $L_s \in [100, 600]$ nm. No improvement in efficiency occurs for $L_s > 600$ nm by the use of a periodically corrugated back reflector. The optimal bandgap of CZTSSe remains the same as with the flat back reflector; also, the optimal corrugation parameters lie in narrow ranges: $L_g \in [99, 105]$ nm, $\zeta \in [0.5, 51]$, and $L_x \in [500, 510]$ nm. The data thus indicate the slight benefit of exciting both surface-plasmon-polariton waves [37–39] and waveguide modes [41, 42] by taking advantage of the grating corrugations [43], but only when the photon-absorbing layer is ultrathin.

4.3   ALGAAS SOLAR CELLS

Gallium arsenide (GaAs) is a III–V semiconductor with a direct bandgap of 1.42 eV. Due to excellent efficiency and a significantly small weight-to-power ratio [65], GaAs solar cells are extensively deployed for extra-terrestrial applications. The high cost of GaAs solar cells is attributed to the expensive materials and substrates [66], with the substrate accounting for at least

Table 4.8: Model-predicted parameters of the optimal MgF$_2$/AZO/od-ZnO/CdS/CZTSSe/Al$_2$O$_3$/Mo solar cell with a specified value of $L_s \in [100, 2200]$ nm, when the CZTSSe layer is homogeneous and the back reflector is periodically corrugated [57]

L_s (nm)	ξ	L_g (nm)	ζ	L_x (nm)	J_{sc} (mA cm^{-2})	V_{oc} (mV)	FF (%)	η (%)
100	0.5172	100	0.50	500	19.99	506	75.2	7.62
200	0.5000	105	0.51	510	25.34	532	72.3	9.75
300	0.5000	100	0.50	500	27.87	546	70.4	10.72
400	0.4828	103	0.51	502	28.56	547	69.7	10.91
600	0.4655	99	0.50	508	29.43	556	70.2	11.50
1200	0.5172	101	0.51	500	30.13	558	70.3	11.84
2200	0.5000	100	0.50	500	30.00	557	70.3	11.76

80% of the total cost of a GaAs solar cell [67]. Different strategies such as wafer reuse and epitaxial lift-off have been investigated to reduce the cost [67]. The record efficiency is 28.8%, reported for thin-film solar cells fabricated using epitaxial lift-off [68]. Even though the epitaxial lift-off procedure reduces material use, GaAs solar cells remain prohibitively expensive for terrestrial applications [69]. Further cost reduction indicates that GaAs should be replaced by a compound semiconductor with GaAs as one of its constituents.

Al$_\xi$Ga$_{1-\xi}$As is an alloy with bandgap energy E$_g$ ranging from 1.42 eV ($\xi = 0$) to 2.168 eV ($\xi = 1$) [70, 71]. The compositional dependence of the AlGaAs bandgap is divided into two regimes, with $\xi \leq 0.45$ for the direct-bandgap regime, $\xi > 0.45$ for the indirect-bandgap regime, and

$$
E_g = \begin{cases} 1.424 + 1.247\xi \text{ eV}, & \xi \leq 0.45, \\ 1.9 + 0.125\xi + 0.143\xi^2 \text{ eV}, & \xi > 0.45. \end{cases} \tag{4.2}
$$

The enhanced bandgap energy of AlGaAs offers considerable hope for higher efficiency.

4.3.1   STRUCTURE

The structure of AlGaAs thin-film solar cell can be described as MgF$_2$/ZnS/FSP/p-type GaAs(AlGaAs)/n-type GaAs(AlGaAs)/BSP/backing-layer [72], as shown in Fig. 4.7. The anti-reflection coating comprises an MgF$_2$ layer on top of a layer of zinc sulfide (ZnS). The front-surface passivation layer is made of either heavily doped aluminum-indium phosphide (AlInP) or heavily doped AlGaAs of p type, this layer also connected to the front electrode. Then comes an emitter layer of either p-type GaAs or p-type AlGaAs, followed by the main photon-absorbing layer made of n-type AlGaAs. The back-surface passivation layer is of either heavily doped gallium-indium phosphide (GaInP) or heavily doped AlGaAs of n type. The back electrode commonly used in thin-film GaAs solar cells is a Pd-Ge-Au trilayer.

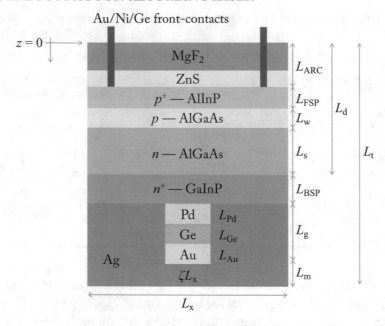

Figure 4.7: Schematic of the reference unit cell of the AlGaAs solar cell with a homogeneous photon-absorbing layer and a custom backing layer.

Passivation plays a vital role in realizing high-efficiency solar cells, especially ultrathin solar cells [73, 74]. As GaInP is lattice-matched to GaAs/AlGaAs, a thin GaInP layer is utilized as the back-surface passivation layer thin-film GaAs solar cells [72]. However, the trilayered Pd-Ge-Au (or Ni-Ge-Au) back electrode used for GaAs solar cells is a poor reflector [72]. Researchers are now investigating a layer of a highly reflective metal (such as silver) in which localized trilayered Pd-Ge-Au (or Ni-Ge-Au) back-contacts are embedded [75], with ratio $\zeta \in (0, 1)$ to be optimized for best performance, as shown in Fig. 4.7.

4.3.2   GEOMETRIC DESCRIPTION

The solar cell occupies the region $\mathcal{X} : \{(x, y, z)| -\infty < x < \infty, -\infty < y < \infty, 0 < z < L_t\}$, with the half spaces $z < 0$ and $z > L_t$ occupied by air. The reference unit cell $\mathcal{R} : \{(x, y, z)| - L_x/2 < x < L_x/2, -\infty < y < \infty, 0 < z < L_t\}$ is schematically depicted in Fig. 4.7, with the materials used by Ahmad et al. [76] for optoelectronic optimization identified therein.

The region $0 < z < L_{ARC}$ is an anti-reflection coating comprising a 110-nm-thick layer of MgF_2 [12] and a 150-nm-thick layer of ZnS [29]. The 20-nm-thick front-surface passivation layer of p^+-$Al_{0.51}In_{0.49}P$ [77] reduces the front-surface recombination rate and thereby improves J_{sc} [74]. Next, homogeneous p-$Al_{\bar{\xi}}Ga_{1-\bar{\xi}}As$ [78] with fixed $\bar{\xi}$ occupies the 50-

nm-thick region $L_{ARC} + L_{FSP} < z < L_{ARC} + L_{FSP} + L_w$ to form a p-n junction with an n-$Al_\xi Ga_{1-\xi} As$ [78] photon-absorbing layer of thickness L_s. The 20-nm-thick layer of n^+-$Ga_{0.49}In_{0.51}P$ [79] is the back-surface passivation layer that reduces the back-surface recombination rate and thereby improves J_{sc} [73]. A Pd-Ge-Au trilayer of total thickness $L_g = 170$ nm and width ζL_x, $\zeta \in (0, 1)$, along the x axis serves as the localized ohmic back-contact [75] comprising a palladium layer of thickness $L_{Pd} = 20$ nm [80], a germanium layer of thickness $L_{Ge} = 50$ nm [81], and a gold layer of thickness $L_{Au} = 100$ nm [80]. The remainder of the region $L_d + L_{BSP} < z < L_d + L_{BSP} + L_g$ is occupied by silver [80] for optical reflection. The solar cell is terminated by a 100-nm-thick layer of silver.

4.3.3 OPTICAL DESCRIPTION

Spectrums of the relative permittivity $\varepsilon/\varepsilon_0$ of MgF_2 [29], ZnS [29], AlInP [77], GaInP [79], Pd [80], Ge [81], Au [80], and Ag [80] are provided in Fig. 4.8 for $\lambda_0 \in [300, 950]$ nm. The real and imaginary parts of the relative permittivity $\varepsilon/\varepsilon_0$ of AlGaAs are provided in Fig. 4.9 as functions of $\lambda_0 \in [300, 950]$ nm and $\xi \in [0, 0.8]$ [78], data for $\xi \in (0.8, 1]$ being unavailable.

4.3.4 ELECTRICAL DESCRIPTION

The region $L_{ARC} < z < L_d + L_{BSP}$ contains four semiconductor layers: the p^+-AlInP FSP layer, the p-AlGaAs layer, the n-AlGaAs absorber layer, and the n^+-GaInP BSP layer. This region must be considered in the electrical submodel [33, 76] because all four layers contribute to charge-carrier generation. The n^+-GaInP back-surface passivation layer in AlGaAs solar cells is a semiconductor in contrast to the Al_2O_3 dielectric layer in CIGS and CZTSSe solar cells. Table 4.9 provides the values of electrical parameters used for all four semiconductors [70, 71, 82–84].

4.3.5 COUPLED OPTOELECTRONIC MODELING AND OPTIMIZATION

The coupled optoelectronic model of Chapters 2 and 3 [33] has been validated by comparison with the experimental results for the MgF_2/ZnS/AlInP/p-GaAs/n-GaAs/GaInP/Pd-Ge-Au solar cell (i.e., $\bar{\xi} = \xi = 0$) described by $L_s = 2000$ nm, $\zeta = 1$, and $L_m = 0$ [76]. Values of J_{sc}, V_{oc}, FF, and η obtained from the model are provided in Table 4.10, as also are the corresponding experimental data [72]. The model predictions are in reasonable agreement with the experimental data. Furthermore, the model-predicted efficiency of 27.4% is close to the highest efficiency (27.6%) reported [68] for GaAs solar cells. No interface defects were taken into consideration in the model, suggesting that all the experimentally observed characteristics can be accounted for [83] by the bulk properties of MgF_2, ZnS, AlInP, p-GaAs, n-GaAs, GaInP, palladium, germanium, and gold.

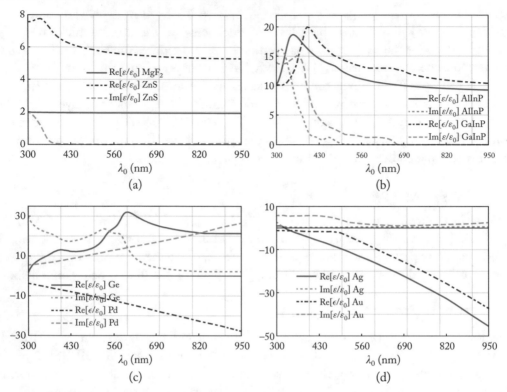

Figure 4.8: Real and imaginary parts of the relative permittivity $\varepsilon/\varepsilon_0$ of (a) MgF_2 and ZnS, (b) AlInP and GaInP, (c) Pd and Ge, and (d) Au and Ag as functions of $\lambda_0 \in [300, 950]$ nm. The imaginary part of the relative permittivity of MgF_2 is negligibly small.

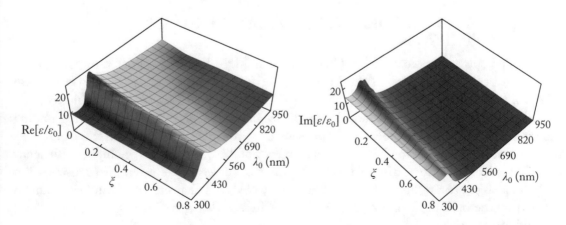

Figure 4.9: Real and imaginary parts of the relative permittivity $\varepsilon/\varepsilon_0$ of AlGaAs as functions of $\lambda_0 \in [300, 950]$ nm and $\xi \in [0, 0.8]$.

Table 4.9: Electrical properties of AlInP, GaInP, and $Al_{\xi}Ga_{1-\xi}As$

Parameter (unit)	AlInP [82, 83]	GaInP [82–84]	$Al_{\xi}Ga_{1-\xi}As$ [70, 71, 84]
E_g (eV)	2.35	1.9	$1.424 + 1.247\xi$, $0 \leq \xi < 0.45$; $1.9 + 0.125\xi + 0.143\xi^2$, $0.45 \leq \xi \leq 1$
χ (eV)	3.78	4.1	$4.07 - 1.1\xi$, $0 \leq \xi < 0.45$; $3.64 - 0.14\xi$, $0.45 \leq \xi \leq 1$
N_c (cm^{-3})	2.5×10^{18}	6.5×10^{17}	$2.5 \times 10^{19}(0.063 + 0.083\xi)^{3/2}$, $0 \leq \xi < 0.45$; $2.5 \times 10^{19}(0.85 - 0.14\xi)^{3/2}$, $0.45 \leq \xi < 1$
N_v (cm^{-3})	7×10^{18}	1.5×10^{19}	$2.5 \times 10^{19}(0.51 + 0.25\xi)^{3/2}$
N_D (cm^{-3})		2×10^{18}	1×10^{18} (n-type AlGaAs)
N_A (cm^{-3})	2×10^{18}		1×10^{18} (p-type AlGaAs)
μ_n (cm^2 V^{-1} s^{-1})	100	500	$8 \times 10^3 - 2.2 \times 10^4\xi + 10^4\xi^2$, $0 \leq \xi < 0.45$; $-255 + 1160\xi - 720\xi^2$, $0.45 \leq \xi \leq 1$
μ_p (cm^2 V^{-1} s^{-1})	10	30	$370 - 970\xi + 740\xi^2$
ε_{dc}	11.8	11.8	$13.18 - 3.12\xi$
N_f (cm^{-3})	10^{17}	10^{17}	$(1 + 9\xi) \times 10^{15}$
E_T (eV)	$0.5E_g$	$0.5E_g$	$E_c - 0.75$ eV
σ_n (cm^2)	10^{-14}	10^{-14}	10^{-16}
σ_p (cm^2)	10^{-14}	10^{-14}	10^{-16}
R_B (cm^3 s^{-1})	10^{-10}	10^{-10}	1.8×10^{-10}
$v_{th,n}$ (cm s^{-1})	10^7	10^7	$(4.4 - 2.1\xi) \times 10^7$
$v_{th,p}$ (cm s^{-1})	10^7	10^7	$(1.8 - 0.5\xi) \times 10^7$
C_n (cm^6 s^{-1})	10^{-30}	10^{-30}	10^{-30}
C_p (cm^6 s^{-1})	10^{-30}	10^{-30}	10^{-30}

Table 4.10: Comparison of J_{sc}, V_{oc}, FF, and η predicted by the coupled optoelectronic model [76] for a GaAs solar cell with a homogeneous n-GaAs photon-absorbing layer with experimental data [72]; $\bar{\xi} = \xi = 0$, $L_s = 2000$ nm, $\zeta = 1$, and $L_m = 0$

	J_{sc} (mA cm^{-2})	V_{oc} (mV)	FF (%)	η (%)
Model [76]	29.8	1081	85.1	27.4
Experiment [72]	29.5	1045	84.6	26.1

Table 4.11: Model-predicted parameters of the optimal AlGaAs solar cell with a specified value of $L_s \in [100, 2000]$ nm, when the n-AlGaAs absorber layer is homogeneous and $L_m = 100$ nm [76]

L_s (nm)	$\bar{\xi}$	ξ	ζ (nm)	L_x	J_{sc} (mA cm^{-2})	V_{oc} (mV)	FF (%)	η (%)
100	0.8	0	0.05	500	18.9	1149	85.1	18.5
200	0.8	0	0.05	510	21.6	1128	85.2	20.7
300	0.8	0	0.05	502	24.1	1124	85.7	23.2
400	0.8	0	0.05	510	25.8	1119	86.5	24.9
500	0.8	0	0.05	500	27.0	1117	86.3	26.1
1000	0.8	0	0.05	500	29.2	1104	87.0	28.1
2000	0.8	0	0.05	510	30.2	1090	87.3	28.8

GaAs Photon-Absorbing Layer

Silver is absent in commercial GaAs solar cells while the Pd-Ge-Au trilayer extends across the entire back surface (i.e., $\zeta = 1$) [68, 72]. The optoelectronic model predicts that localized ohmic back-contacts inserted periodically in a silver layer will enhance efficiency [76], in accord with experimental findings [75]. The optimal parameters predicted are $L_x = 510$ nm and $\zeta = 0.05$. The effect of localized ohmic back-contacts on η is markedly significant for thin photon-absorbing layers but less pronounced for thick ones.

AlGaAs Photon-Absorbing Layer

Optoelectronic optimization of the solar cell with a homogeneous n-AlGaAs absorber layer, a silver back reflector, and localized ohmic Pd-Ge-Au back-contacts has been performed for specified values of $L_s \in [100, 2000]$ nm [76]. Whereas $L_m = 100$ nm was fixed, the four-dimensional (i.e., $\tilde{N} = 4$) parameter space for optimizing η was chosen as: $\bar{\xi} \in [0, 0.8]$, $\xi \in [0, 0.8]$, $L_x \in [100, 1000]$ nm, and $\zeta \in [0.05, 1]$. Values of J_{sc}, V_{oc}, FF, and η predicted by the coupled optoelectronic model are presented in Table 4.11 for seven different values of L_s. The values of $\bar{\xi}$, ξ, L_x, and ζ for the optimal designs are also provided in the same table.

There is remarkable similarity in the optimal design, regardless of the thickness L_s of the photon-absorbing layer. The best results are obtained with $\bar{\xi} = 0.8$ so that p-type semiconductor for the p-n junction is AlGaAs, $\xi = 0$ so that the photon-absorbing layer is made of n-GaAs, $L_x \approx 505$ nm, and $\zeta = 0.05$ [76]. Even lower values of ζ yield higher efficiencies, but the localized ohmic Pd-Ge-Au back-contacts are necessary because of superior electron-collection capability [75]. The efficiency decreases slightly for $\bar{\xi} = 0$, but increases monotonically with increasing L_s because more photons are absorbed in a thicker photon-absorbing layer. These results clearly indicate that, most likely, the only way to improve the efficiency of a GaAs solar cell is by grading the bandgap of the photon-absorbing layer.

4.4 BIBLIOGRAPHY

[1] M. A. Contreras, L. M. Mansfield, B. Egaas, J. Li, M. Romero, R. Noufi, E. Rudiger-Voigt, and W. Mannstadt, Wide bandgap Cu(In,Ga)Se$_2$ solar cells with improved energy conversion efficiency, *Progress in Photovoltaics: Research and Applications*, 20:843–850 (2012). 55

[2] P. Jackson, R. Wuerz, D. Hariskos, E. Lotter, W. Witte, and M. Powalla, Effects of heavy alkali elements in Cu(In,Ga)Se$_2$ solar cells with efficiencies up to 22.6%, *Physica Status Solidi RRL*, 10:583–586 (2016). 55, 56, 59, 61, 62

[3] M. A. Green, Y. Hishikawa, E. D. Dunlop, D. H. Levi, J. Hohl-Ebinger, and A. W. Y. Ho-Baillie, Solar cell efficiency tables (version 51), *Progress in Photovoltaics: Research and Applications*, 26:3–12 (2018). 55, 62

[4] C. Candelise, M. Winskel, and R. Gross, Implications for CdTe and CIGS technologies production costs of indium and tellurium scarcity, *Progress in Photovoltaics: Research and Applications*, 20:816–831 (2012). 55, 62

[5] J. Pettersson, T. Torndahl, C. Platzer-Björkman, A. Hultqvist, and M. Edoff, The influence of absorber thickness on Cu(In,Ga)Se$_2$ solar cells with different buffer layers, *IEEE Journal of Photovoltaics*, 3:1376–1382 (2013). 55

[6] E. Jarzembowski, M. Maiberg, F. Obereigner, K. Kaufmann, S. Krause, and R. Scheer, Optical and electrical characterization of Cu(In,Ga)Se$_2$ thin film solar cells with varied absorber layer thickness, *Thin Solid Films*, 576:75–80 (2015). 55

[7] M. Gloeckler and J. R. Sites, Potential of submicrometer thickness Cu(In,Ga)Se$_2$ solar cells, *Journal of Applied Physics*, 98:103703 (2005). 55, 87

[8] M. Schmid, Review on light management by nanostructures in chalcopyrite solar cells, *Semiconductor Science and Technology*, 32:043003 (2017). 55, 57

[9] E. Jarzembowski, B. Fuhrmann, H. Leipner, W. Fränzel, and R. Scheer, Ultrathin Cu(In,Ga)Se$_2$ solar cells with point-like back contact in experiment and simulation, *Thin Solid Films*, 519:61–65 (2016). 55

[10] C. van Lare, G. Yin, A. Polman, and M. Schmid, Light coupling and trapping in ultrathin Cu(In,Ga)Se$_2$ solar cells using dielectric scattering patterns, *ACS Nano*, 9:9603–9613 (2015). 55

[11] F. Ahmad, T. H. Anderson, P. B. Monk, and A. Lakhtakia, Efficiency enhancement of ultrathin CIGS solar cells by optimal bandgap grading, *Applied Optics*, 58:6067–6078 (2019). 55, 56, 57, 58, 59, 60, 61, 62

F. Ahmad, T. H. Anderson, P. B. Monk, and A. Lakhtakia, Efficiency enhancement of ultrathin CIGS solar cells by optimal bandgap grading, *Applied Optics*, 59:2615 (2020) (erratum).

[12] M. J. Dodge, Refractive properties of magnesium fluoride, *Applied Optics*, 23:1980–1985 (1984). 56, 58, 64, 70

[13] N. Ehrmann and R. Reineke-Koch, Ellipsometric studies on ZnO:Al thin films: Refinement of dispersion theories, *Thin Solid Films*, 519:1475–1485 (2010). 56, 58, 64

[14] J. S. Wellings, A. P. Samantilleke, P. Warren, S. N. Heavens, and I. M. Dharmadasa, Comparison of electrodeposited and sputtered intrinsic and aluminium-doped zinc oxide thin films, *Semiconductor Science and Technology*, 23:125003 (2008). 56, 57, 64

[15] R. E. Treharne, A. Seymour-Pierce, K. Durose, K. Hutchings, S. Roncallo, and D. Lane, Optical design and fabrication of fully sputtered CdTe/CdS solar cells, *Journal of Physics: Conference Series*, 286:012038 (2011). 56, 58, 64

[16] R. Boidin, T. Halenkovič, V. Nazabal, L. Beneš, and P. Němec, Pulsed laser deposited alumina thin films, *Ceramics International*, 42:1177–1182 (2016). 56, 57, 58, 64

[17] D. Rudmann, D. Brémaud, A. F. da Cunha, G. Bilger, A. Strohm, M. Kaelin, H. Zogg, and A. Tiwari, Sodium incorporation strategies for CIGS growth at different temperatures, *Thin Solid Films*, 480–481:55–60 (2005). 56

[18] J. Guillemoles, T. Haalboom, T. Gödecke, F. Ernst, and D. Cahen, Phase and interface stability issues in chalcopyrite-based thin film solar cells, *Material Research Society Symposium Proceedings*, 485:127–132 (1998). 56

[19] C. Frisk, C. Platzer-Björkman, J. Olsson, P. Szaniawski, J. T. Wätjen, V. Fjällström, P. Salomé, and M. Edoff, Optimizing Ga-profiles for highly efficient Cu(In, Ga)Se$_2$ thin film solar cells in simple and complex defect models, *Journal of Physics D: Applied Physics*, 47:485104 (2014). 56, 59, 60, 65

[20] J. Song, S. S. Li, C. H. Huang, O. D. Crisalle, and T. J. Anderson, Device modeling and simulation of the performance of Cu(In$_{1-x}$,Ga$_x$)Se$_2$ solar cells, *Solid-State Electronics*, 48:73–79 (2004). 56, 63

[21] D. Kuciauskas, J. V. Li, M. A. Contreras, J. Pankow, P. Dippo, M. Young, L. M. Mansfield, R. Noufi, and D. Levi, Charge carrier dynamics and recombination in graded band gap CuIn$_{1-x}$Ga$_x$Se$_2$ polycrystalline thin-film photovoltaic solar cell absorbers, *Journal of Applied Physics*, 144:154505 (2013). 56

[22] B. Vermang, J. T. Wätjen, V. Fjällström, F. Rostvall, M. Edoff, R. Kotipalli, F. Henry, and D. Flandre, Employing Si solar cell technology to increase efficiency of ultra-thin Cu(In,Ga)Se$_2$ solar cells, *Progress in Photovoltaics: Research and Applications*, 22:1023–1029 (2014). 57

[23] B. Vermang, V. Fjällström, J. Pettersson, P. Salomé, and M. Edoff, Development of rear surface passivated Cu(In,Ga)Se$_2$ thin film solar cells with nano-sized local rear point contacts, *Solar Energy Materials and Solar Cells*, 117:505–511 (2013). 57

[24] B. Vermang, V. Fjällström, X. Gao, and M. Edoff, Improved rear surface passivation of Cu(In,Ga)Se$_2$ solar cells: A combination of an Al$_2$O$_3$ rear surface passivation layer and nano-sized local rear point contacts, *IEEE Journal of Photovoltaics*, 4:486–492 (2014). 57, 63, 64

[25] P. Casper, R. Hünig, G. Gomard, O. Kiowski, C. Reitz, U. Lemmer, M. Powalla, and M. Hetterich, Optoelectrical improvement of ultra-thin Cu(In,Ga)Se$_2$ solar cells through microstructured MgF$_2$ and Al$_2$O$_3$ back contact passivation layer, *Physica Status Solidi RRL*, 10:376–380 (2016). 57

[26] K. Orgassa, H. W. Schock, and J. H. Werner, Alternative back contact materials for thin film Cu(In,Ga)Se$_2$ solar cells, *Thin Solid Films*, 431–432:387–391 (2003). 57

[27] A. H. Jahagirdar, A. A. Kadam, and N. G. Dhere, Role of i-ZnO in optimizing open circuit voltage of CIGS2 and CIGS thin film solar cells, *Proc. of 4th IEEE World Conference on Photovoltaic Energy*, Waikoloa, HI, May 7–12, 2006. 57

[28] M. F. Iskander, *Electromagnetic Fields and Waves* (Waveland Press, Long Grove, IL, 2012). 57, 64

[29] M. R. Querry, Optical constants of minerals and other materials from the millimeter to the ultraviolet, *Contractor Report CRDEC-CR-88009* (1987). 57, 58, 64, 70, 71

[30] C. Stelling, C. R. Singh, M. Karg, T. A. F. König, M. Thelakkat, and M. Retsch, Plasmonic nanomeshes: Their ambivalent role as transparent electrodes in organic solar cells, *Scientific Reports*, 7:42530 (2017). 58, 64

[31] S. Minoura, T. Maekawa, K. Kodera, A. Nakane, S. Niki, and H. Fujiwara, Optical constants of $Cu(In,Ga)Se_2$ for arbitrary Cu and Ga compositions, *Journal of Applied Physics*, 117:195703 (2015). 58

[32] F. Ahmad, Optoelectronic modeling and optimization of graded-bandgap thin-film solar cells, Ph.D. Dissertation (The Pennsylvania State University, University Park, PA, 2020). 58, 64

[33] T. H. Anderson, B. J. Civiletti, P. B. Monk, and A. Lakhtakia, Coupled optoelectronic simulation and optimization of thin-film photovoltaic solar cells, *Journal of Computational Physics*, 407:109242 (2020). 59, 66, 71

T. H. Anderson, B. J. Civiletti, P. B. Monk, and A. Lakhtakia, Coupled optoelectronic simulation and optimization of thin-film photovoltaic solar cells, *Journal of Computational Physics*, 418:109561 (2020) (corrigendum).

[34] J. AbuShama, R. Noufi, S. Johnston, S. Ward, and X. Wu, Improved performance in $CuInSe_2$ and surface-modified $CuGaSe_2$ solar cells, *Proc. 31st IEEE Photovoltaic Specialists Conference (PVSC)*, pages 299–302, Lake Buena Vista, FL, June 3–7, 2005. 59, 61

[35] S. J. Fonash, *Solar Cell Device Physics*, 2nd ed. (Academic Press, Burlington, MA, 2010). 59

[36] J. Nelson, *The Physics of Solar Cells* (Imperial College Press, London, UK, 2003). 59

[37] L. M. Anderson, Harnessing surface plasmons for solar energy conversion, *Proc. of SPIE*, 408:172–178 (1983). 60, 68

[38] C. Heine and R. H. Morf, Submicrometer gratings for solar energy applications, *Applied Optics*, 34:2476–2482 (1995). 60, 68

[39] H. A. Atwater and A. Polman, Plasmonics for improved photovoltaic devices, *Nature Materials*, 9:205–213 (2010). 60, 68

[40] A. S. Hall, M. Faryad, G. D. Barber, L. Liu, S. Erten, T. S. Mayer, A. Lakhtakia, and T. E. Mallouk, Broadband light absorption with multiple surface plasmon polariton waves excited at the interface of a metallic grating and photonic crystal, *ACS Nano*, 7:4995–5007 (2013). 60

[41] F.-J. Haug, K. Söderström, A. Naqavi, and C. Ballif, Excitation of guided-mode resonances in thin film silicon solar cells, *Materials Research Society Proceedings*, 1321:123–128 (2011). 60, 68

[42] T. Khaleque and R. Magnusson, Light management through guided-mode resonances in thin-film silicon solar cells, *Journal of Nanophotonics*, 8:083995 (2014). 60, 68

[43] J. A. Polo Jr., T. G. Mackay, and A. Lakhtakia, *Electromagnetic Surface Waves: A Modern Perspective* (Elsevier, Waltham, MA, 2013). 60, 68

[44] S. Adachi, *Earth-Abundant Materials for Solar Cell* (Wiley, Chichester, West Sussex, UK, 2015). 62, 64, 65, 67

[45] W. Wang, M. T. Winkler, O. Gunawan, T. Gokmen, T. K. Todorov, Y. Zhu, and D. B. Mitzi, Device characteristics of CZTSSe thin-film solar cells with 12.6% efficiency, *Advanced Energy Materials*, 4:1301465 (2014). 62, 63, 66, 67

[46] L. H. Wong, A. Zakutayev, J. D. Major, X. Hao, A. Walsh, T. K. Todorov, and E. Saucedo, Emerging inorganic solar cell efficiency tables (Version 1), *Journal of Physics: Energy*, 1:032001 (2019). 62

[47] T. Gokmen, O. Gunawan, T. K. Todorov, and D. B. Mitzi, Band tailing and efficiency limitation in kesterite solar cells, *Applied Physics Letters*, 103:103506 (2013). 62, 65

[48] C. Frisk, T. Ericson, S.-Y. Li, P. Szaniawski, J. Olsson, and C. Platzer-Björkman, Combining strong interface recombination with bandgap narrowing and short diffusion length in Cu_2ZnSnS_4 device modeling, *Solar Energy Materials and Solar Cells*, 144:364–370 (2016). 62, 65, 66, 67

[49] I. L. Repins, H. Moutinho, S. G. Choi, A. Kanevce, D. Kuciauskas, P. Dippo, C. L. Beall, J. Carapella, C. DeHart, B. Huang, and S. H. Wei, Indications of short minority-carrier lifetime in kesterite solar cells, *Journal of Applied Physics*, 114:084507 (2013). 62

[50] O. Gunawan, T. K. Todorov, and D. B. Mitzi, Loss mechanisms in hydrazine-processed $Cu_2ZnSn(S,Se)_4$ solar cells, *Applied Physics Letters*, 97:233506 (2010). 62

[51] T. Gokmen, O. Gunawan, and D. B. Mitzi, Minority carrier diffusion length extraction in $Cu_2ZnSn(Se,S)_4$ solar cells, *Journal of Applied Physics*, 114:114511 (2013). 62, 68

[52] T. Gershon, T. Gokmen, O. Gunawan, R. Haight, S. Guha, and B. Shin, Understanding the relationship between $Cu_2ZnSn(S,Se)_4$ material properties and device performance, *MRS Communications*, 4:159–170 (2014). 62

[53] D. B. Mitzi, O. Gunawan, T. K. Todorov, K. Wang, and S. Guha, The path towards a high-performance solution-processed kesterite solar cell, *Solar Energy Materials and Solar Cells*, 95:1421–1436 (2011). 62, 63, 66, 67

[54] A. Kanevce, I. L. Repins, and S. H. Wei, Impact of bulk properties and local secondary phases on the $Cu_2ZnSn(S,Se)_4$ solar cells open-circuit voltage, *Solar Energy Materials and Solar Cells*, 133:119–125 (2015). 63, 65, 67

[55] Y. S. Lee, T. Gershon, O. Gunawan, T. K. Todorov, T. Gokman, Y. Virgus, and S. Guha, $Cu_2ZnSnSe_4$ thin-film solar cells by thermal co-evaporation with 11.6% efficiency and improved minority carrier diffusion length, *Advanced Energy Materials*, 5:1401372 (2015). 63

[56] S. Ahmed, K. B. Reuter, O. Gunawan, L. Guo, L. T. Romankiw, and H. Deligianni, A high efficiency electrodeposited Cu_2ZnSnS_4 solar cell, *Advanced Energy Materials*, 6:253–259 (2012). 63

[57] F. Ahmad, A. Lakhtakia, T. H. Anderson, and P. B. Monk, Towards highly efficient thin-film solar cells with a graded-bandgap CZTSSe layer, *Journal of Physics: Energy*, 2:025004 (2020). 63, 64, 66, 67, 68, 69

F. Ahmad, A. Lakhtakia, T. H. Anderson, and P. B. Monk, Towards highly efficient thin-film solar cells with a graded-bandgap CZTSSe layer, *Journal of Physics: Energy*, 2:039501 (2020) (corrigendum).

[58] F. Liu, J. Huang, K. Sun, C. Yan, Y. Shen, J. Park, A. Pu, F. Zhou, X. Liu, J. A. Stride, M. A. Green, and X. Hao, Beyond 8% ultrathin kesterite Cu_2ZnSnS_4 solar cells by interface reaction route controlling and self-organized nanopattern at the back contact, *NPG Asia Materials*, 9:e401 (2017). 63, 64

[59] G. Altamura, L. Grenet, C. Roger, F. Roux, V. Reita, R. Fillon, H. Fournier, S. Perraud, and H. Mariette, Alternative back contacts in kesterite $Cu_2ZnSn(S_{1-x}Se_x)_4$ thin film solar cells, *Journal of Renewable and Sustainable Energy*, 6:011401 (2014). 64

[60] Y. Hirate, H. Tampo, S. Minoura, H. Kadowaki, A. Nakane, K. M. Kim, H. Shibata, S. Niki, and H. Fujiwara, Dielectric functions of $Cu_2ZnSnSe_4$ and Cu_2SnSe_3 semiconductors, *Journal of Applied Physics*, 117:015702 (2015). 64

[61] A. Nakane, H. Tampo, M. Tamakoshi, S. Fujimoto, K. M. Kim, S. Kim, H. Shibata, S. Niki, and H. Fujiwara, Quantitative determination of optical and recombination losses in thin-film photovoltaic devices based on external quantum efficiency analysis, *Journal of Applied Physics*, 120:064505 (2016). 64

[62] A. R. Beal and H. P. Hughes, Kramers–Krönig analysis of the reflectivity spectra of 2H-MoS_2, 2H-$MoSe_2$, and 2H-$MoTe_2$, *Journal of Physics C: Solid State Physics*, 12:881–890 (1979). 66

[63] T. Gokmen, O. Gunawan, and D. B. Mitzi, Semi-empirical device model for $Cu_2ZnSn(S,Se)_4$ solar cells, *Applied Physics Letters*, 105:033903 (2014). 66

[64] W. Shockley and H. J. Queisser, Detailed balance limit of efficiency of p-n junction solar cells, *Journal of Applied Physics*, 32:510–519 (1961). 67

[65] A. van Geelen, P. R. Hageman, G. J. Bauhuis, P. C. van Rijsingen, P. Schmidt, and L. J. Giling, Epitaxial lift-off GaAs solar cell from a reusable GaAs substrate, *Materials Science and Engineering B*, 45:162–171 (1997). 68

[66] J. P. Connolly, D. Mencaraglia, C. Renard, and D. Bouchier, Designing III-V multijunction solar cells on silicon, *Progress in Photovoltaics: Research and Applications*, 22:810–820 (2013). 68

[67] J. S. Ward, T. Remo, K. Horowitz, M. Woodhouse, B. Sopori, K. VanSant, and P. Basore, Techno-economic analysis of three different substrate removal and reuse strategies for III-V solar cells, *Progress in Photovoltaics: Research and Applications*, 24:1284–1292 (2016). 69

[68] B. M. Kayes, H. Nie, R. Twist, S. G. Spruytte, F. Reinhardt, I. C. Kizilyalli, and G. S. Higashi, 27.6% conversion efficiency, a new record for single-junction solar cells under 1 sun illumination, *Proc. of 37th IEEE Photovoltaic Specialists Conference (PVSC)*, pages 4–8, Seattle, WA, June 19–24, 2011. 69, 71, 74

[69] K. A. W. Horowitz, T. Remo, B. Smith, and A. Ptak, Techno-economic analysis and cost reduction roadmap for III-V solar cells, *NREL Technical Report NREL/TP-6A20-72103* (2018). 69

[70] S. Adachi, GaAs, AlAs, and $Al_x Ga_{1-x}$As: Material parameters for use in research and device applications, *Journal of Applied Physics*, 58:R1–R29, (1985). 69, 71, 73

[71] S. Adachi, Ed., *Properties of Aluminum Gallium Arsenide*, EMIS Datareviews Series no. 7 INSPEC (Institution of Electrical Engineers, London, UK, 1993). 69, 71, 73

[72] G. Bauhuis, P. Mulder, E. J. Haverkamp, J. C. C. M. Huijben, and J. J. Schermer, 26.1% thin-film GaAs solar cell using epitaxial lift-off, *Solar Energy Materials and Solar Cells*, 93:1488–1491 (2009). 69, 70, 71, 74

[73] O. von Roos, A simple theory of back surface field (BSF) solar cells, *Journal of Applied Physics*, 49:3503–3511 (1978). 70, 71

[74] S. R. Kurtz, J. M. Olson, D. J. Friedman, J. F. Geisz, K. A. Bertness, and A. E. Kibbler, Passivation of interfaces in high-efficiency photovoltaic devices, *Materials Research Society Proceedings*, 573:95–106 (1999). 70

[75] N. Vandamme, H.-L. Chen, A. Gaucher, B. Behaghel, A. Lemaître, A. Cattoni, C. Dupuis, N. Bardou, J. F. Guillemoles, and S. Collin, Ultrathin GaAs solar cells with a silver back mirror, *IEEE Journal of Photovoltaics*, 5:565–570 (2015). 70, 71, 74, 75

[76] F. Ahmad, A. Lakhtakia, and P. B. Monk, Optoelectronic optimization of graded-bandgap thin-film AlGaAs solar cells, *Applied Optics*, 59:1018–1027 (2020). 70, 71, 74, 75

[77] E. Ochoa-Martínez, L. Barrutia, M. Ochoa, E. Barrigón, I. García, I. Rey-Stolle, C. Algora, P. Basa, G. Kronome, and M. Gabás, Refractive indexes and extinction coefficients of n- and p-type doped GaInP, AlInP, and AlGaInP for multijunction solar cells, *Solar Energy Materials and Solar Cells*, 174:388–396 (2018). 70, 71

[78] D. E. Aspnes, S. M. Kelso, R. A. Logan, and R. Bhat, Optical properties of $Al_xGa_{1-x}As$, *Journal of Applied Physics*, 60:754–767 (1986). 70, 71

[79] M. Schubert, V. Gottschalch, C. M. Herzinger, H. Yao, P. G. Snyder, and J. A. Woollam, Optical constants of $Ga_xIn_{1-x}P$ lattice matched to GaAs, *Journal of Applied Physics*, 77:3416–3419 (1995). 71

[80] P. B. Johnson and R. W. Christy, Optical constants of the noble metals, *Physical Review B*, 6:4370–4379 (1972). 71

[81] G. E. Jellison Jr., Optical functions of GaAs, GaP, and Ge determined by two-channel polarization modulation ellipsometry, *Optical Materials*, 1:151–160 (1992). 71

[82] I. Vurgaftman, J. R. Meyer, and L. R. Ram-Mohan, Band parameters for III–V compound semiconductors and their alloys, *Journal of Applied Physics*, 89:5815–5875 (2001). 71, 73

[83] A. S. Gudovskikh, N. A. Kaluzhniy, V. M. Lantratov, S. A. Mintairov, M. Z. Shvarts, and V. M. Andreev, Numerical modelling of GaInP solar cells with AlInP and AlGaAs windows, *Thin Solid Films*, 516:6739–6743 (2008). 71, 73

[84] Ioffe Institute, $Al_xGa_{1-x}As$ (accessed May 17, 2021). 71, 73

CHAPTER 5

Linearly Graded Photon-Absorbing Layer

The coupled optoelectronic model of Chapters 2 and 3 [1] has been validated in Chapter 4 against experimental data for thin-film CIGS, CZTSSe, and AlGaAs solar cells with homogeneous photon-absorbing layers. Model-predicted values of the open-circuit voltage V_{oc}, short-circuit current density J_{sc}, fill factor FF, and power-conversion efficiency η are in reasonable agreement with their experimental counterparts [2–4]. Also, model-predicted results in Chapter 4 indicate the slight benefit of exciting both surface-plasmon-polariton waves [5–7] and waveguide modes [8, 9] by taking advantage of the grating corrugations [10], but only when the photon-absorbing layer is ultrathin. Front-surface and back-surface passivation layers reduce the electron-hole-pair recombination rate R and help enhance the efficiency to some extent.

Any significant improvement of efficiency is likely to come from altering the photon-absorbing layer, but it would not suffice to merely increase the optical short-circuit current density J_{sc}^{opt} because R too must be kept as low as possible. Clearly then, optical analysis of thin-film solar cells by itself is inadequate for solar-cell design [11, 12]. More importantly, recourse needs to be taken to grading the bandgap of the photon-absorbing layer, as suggested by optoelectronic modeling of thin-film Schottky-barrier solar cells with a periodically graded photon-absorbing layer of indium-gallium nitride [13].

The bandgap energy E_g is a function of $\xi \in [0, 1]$, the parameter which quantifies the gallium-indium ratio in CIGS, the sulfur-selenium ratio in CZTSSe, and the aluminum-gallium ratio in AlGaAs. Therefore, the bandgap of the photon-absorbing layer can be graded in the thickness direction by changing ξ dynamically during fabrication [14–16]. Bandgap grading can increase V_{oc} and therefore η, as has been experimentally established for CZTSSe [15, 17, 18] and AlGaAs [16] solar cells. Bandgap grading creates a drift electric field that accelerates minority charge-carriers in that layer toward the p-n junction [14, 19]. Also, bandgap grading can allow photon absorption over a wider spectral regime. Efficiency increase by bandgap grading has also been predicted by simple simulations of CIGS [20, 21] and CZTSSe [22–25] solar cells. However, simple simulations as well as experiments have shown that linear grading of the bandgap can significantly reduce J_{sc} in CIGS solar cells [14, 20, 21, 26], whereas the efficiency gain in CZTSSe solar cells can be quite modest. Danger lies in suboptimal bandgap grading, as that can reduce J_{sc} to offset the increase in V_{oc}.

5.1   LINEAR GRADING PROFILES

The bandgap profile of the photon-absorbing layer and the dimensions of the back reflector have been optimized for CIGS [2], CZTSSe [3], and AlGaAs [4] solar cells using the coupled optoelectronic model and the differential evolution algorithm [28, 29], as discussed next in this chapter for linear grading of the bandgap energy. All optical and electrical parameters needed are provided in Chapter 4.

Schematics of the three solar cells are available in Figs. 4.1, 4.4, and 4.7. The linearly graded bandgap energy for forward grading of the photon-absorbing layer is given as

$$E_g(z) = E_a + A\left(E_b - E_a\right)\left[\frac{z - (L_d - L_s)}{L_s}\right], \quad z \in (L_d - L_s, L_d), \tag{5.1}$$

where E_a is the minimum bandgap energy, E_b is the maximum bandgap energy, and A is an amplitude (with $A = 0$ representing a homogeneous layer as in Chapter 4). The bandgap energy is smaller near the front interface $z = L_d - L_s$ than near the back interface $z = L_d$.

The linearly graded bandgap energy for backward grading of the photon-absorbing layer is given as

$$E_g(z) = E_b - A\left(E_b - E_a\right)\left[\frac{z - (L_d - L_s)}{L_s}\right], \quad z \in (L_d - L_s, L_d). \tag{5.2}$$

The bandgap energy is larger near the front interface $z = L_d - L_s$ than near the back interface $z = L_d$.

5.2   CIGS SOLAR CELLS

A schematic of the CIGS solar cell is available in Fig. 5.1. The six-dimensional (i.e., $\tilde{N} = 6$) parameter space for maximizing the efficiency of a CIGS solar cell with a linearly bandgap-graded CIGS layer of a specified thickness $L_s \in [100, 2200]$ nm is as follows: $E_a \in [0.947, 1.626]$ eV, $E_b \in [0.947, 1.626]$ eV, $A \in [0, 1]$, $L_g \in [1, 550]$ nm, $\zeta \in (0, 1)$, and $L_x \in [100, 1000]$ nm. Optimization must be carried out subject to the constraint $E_b \geq E_a$. The parameter space is four-dimensional when the back reflector is flat, because $L_g = 0$ and ζ becomes irrelevant.

5.2.1   FORWARD GRADING

Optoelectronic optimization yields $A = 0$ when Eq. (5.1) describes the grading profile [2]. Accordingly, Table 4.3 holds when the back reflector is flat and Table 4.4 when the back reflector is periodically corrugated. Thus, forward grading of the photon-absorbing layer of a CIGS solar cell is predicted by the coupled optoelectronic model [1] to be suboptimal.

5.2.2   BACKWARD GRADING

When Eq. (5.1) is replaced by Eq. (5.2), optoelectronic optimization predicts $A \simeq 1$ for maximum efficiency [2]. Table 5.1 presents η for nine representative values of L_s, when the back

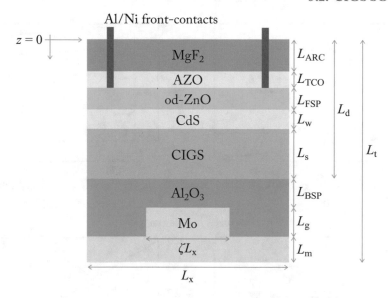

Figure 5.1: Schematic of the reference unit cell of the CIGS solar cell with a linearly graded photon-absorbing layer and a periodically corrugated back reflector. The corrugation height $L_g = 0$ for a flat back reflector.

reflector is periodically corrugated. The corrugations increase the efficiency slightly when $L_s \leq$ 600 nm but are ineffective for thicker CIGS layers.

For $L_s = 100$ nm, the optimal $\eta = 9.88\%$ in Table 5.1, whereas $\eta = 7.25\%$ in Table 4.4. This relative enhancement of 36.27% must be attributed to the backward grading of the bandgap energy of the CIGS layer. Concurrently, J_{sc} increases slightly from 14.89 mA cm^{-2} to 15.09 mA cm^{-2} (1.3% relative increase) but V_{oc} increases significantly from 624 mV to 960 mV (53.84% relative increase); however, the fill factor reduces from 78% to 68%.

For higher values of L_s, the overall trend for backward grading encompasses the enhancement of both V_{oc} and η as well as the reduction of both J_{sc} and FF. For $L_s = 600$ nm, the optimal efficiency increases from 13.79% (Table 4.4) by 14.5% to 15.79% (Table 5.1) and V_{oc} increases from 704 mV by 45.31% to 1023 mV, but J_{sc} decreases from 24.66 mA cm^{-2} to 21.48 mA cm^{-2} and FF reduces from 79% to 71%. Higher values of V_{oc} are positively correlated with larger values of $E_g(z)$ in the photon-absorbing layer near the p-n junction.

Although the relative enhancement in efficiency due to backward grading decreases as L_s increases, the coupled optoelectronic model clearly delivers an important result: *bandgap grading of the photon-absorbing layer can significantly improve η.*

The coupled optoelectronic model yields detailed information about the solar cell. As an example, Fig. 5.2 presents the variations of E_g, χ, E_c, E_v, E_i, n, p, n_i, G, R, and N_f with z inside the optimal CIGS solar cell with a 600-nm-thick backward-graded CIGS layer, the bandgap

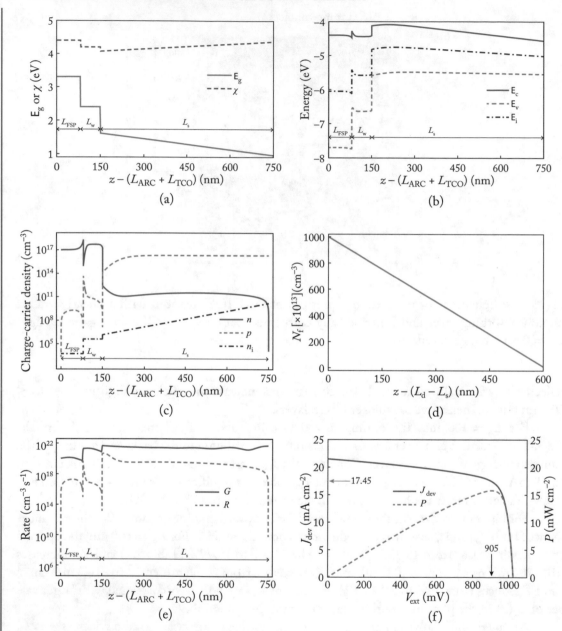

Figure 5.2: Variations of (a) E_g and χ; (b) E_c, E_v, and E_i; (c) n, p, and n_i; (d) N_f; and (e) G and R with z in the optimal CIGS solar cell with a 600-nm-thick backward-graded CIGS photon-absorbing layer. (f) Plots of J_{dev} and P vs. V_{ext}. Values of J_{dev} and V_{ext} for maximum P are identified. The bandgap and geometric parameters are available in Table 5.1.

Table 5.1: Model-predicted parameters of the optimal CIGS solar cell with a specified value of $L_s \in [100, 2200]$ nm, when the CIGS layer is backward graded according to Eq. (5.2) and the back reflector is periodically corrugated [2]

L_s (nm)	E_a (eV)	E_b (eV)	A	L_g (nm)	ζ	L_x (nm)	J_{sc} (mA cm^{-2})	V_{oc} (mV)	FF (%)	η (%)
100	0.96	1.61	0.98	97	0.50	500	15.09	960	68	9.88
200	0.96	1.61	0.98	101	0.51	510	19.28	995	62	12.08
300	0.96	1.62	0.98	101	0.49	502	20.70	1010	63	13.34
400	0.96	1.62	0.98	101	0.49	510	21.13	1011	67	14.34
500	0.95	1.62	0.99	106	0.48	510	21.31	1017	69	15.14
600	0.95	1.62	0.98	101	0.50	500	21.48	1023	71	15.79
900	0.95	1.62	0.99	101	0.50	500	22.21	1032	75	17.24
1200	0.95	1.62	0.99	101	0.50	500	22.74	1037	76	18.07
2200	0.95	1.62	0.98	101	0.50	500	24.09	1039	77	19.27

and geometric parameters being available in Table 5.1. The spatial variations of E_c and E_i are similar to that of E_g and provide the conditions to enhance the generation rate [7]. The intrinsic carrier density n_i varies linearly such that it is small where E_g is large and vice versa. The electron-hole-pair generation rate G is higher near the front face $z = L_d - L_s$ and the back face $z = L_d$ of the CIGS layer and slightly lower in the middle of that layer, but the electron-hole-pair recombination rate R drops sharply near the back face of that layer. Furthermore, the spatial profile of R follows that of the defect density N_f. The J_{dev}–V_{ext} and P–V_{ext} characteristics of the solar cell are shown in this figure, which yields $J_{dev} = 17.45$ mA cm^{-2} and $V_{ext} = 905$ mV for best performance along with FF $= 71\%$ and $\eta = 15.79\%$.

5.3   CZTSSE SOLAR CELLS

A schematic of the CZTSSe solar cell is provided in Fig. 5.3. Equations (5.1) and (5.2) hold for the optical submodel, but $E_g(z)$ should be replaced by $E_g(z) - 0.14\xi(z)$ for the electrical submodel to account for bandtail states, as explained in Section 4.2.4. The six-dimensional (i.e., $\tilde{N} = 6$) parameter space for maximizing the efficiency of a CZTSSe solar cell with a linearly bandgap-graded CZTSSe layer of a specified thickness $L_s \in [100, 2200]$ nm is: $E_a \in [0.91, 1.49]$ eV, $E_b \in [0.91, 1.49]$ eV, $A \in [0, 1]$, $L_g \in [1, 550]$ nm, $\zeta \in (0, 1)$, and $L_x \in [100, 1000]$ nm, subject to the constraint $E_b \geq E_a$. When the back reflector is flat, $\tilde{N} = 4$ because $L_g = 0$ and ζ becomes irrelevant.

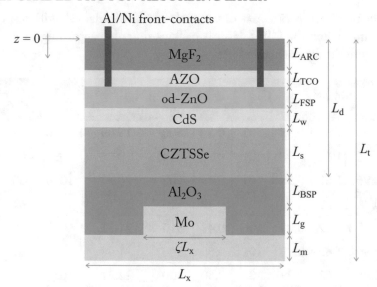

Figure 5.3: Schematic of the reference unit cell of the CZTSSe solar cell with a linearly graded photon-absorbing layer and a periodically corrugated back reflector. The corrugation height $L_g = 0$ for a flat back reflector.

5.3.1   FORWARD GRADING

When Eq. (5.1) holds, the bandgap energy in the CZTSSe layer is smaller near the *p-n* junction than near the back interface for $A > 0$. Table 5.2 presents η for seven representative values of L_s when the back reflector is flat, and Table 5.3 when the back reflector is periodically corrugated [3]. Bandgap and geometric parameters for the optimal designs are also provided in both tables.

A comparison of these two tables shows that periodic corrugation of the back reflector slightly improves η for $L_s \lesssim 600$ nm. Thus, for $L_s = 200$ nm, the maximum efficiency predicted is 11.04% with a flat back reflector and 11.69% with a periodically corrugated back reflector. Whether the back reflector is flat or periodically corrugated, the optimal parameters for forward grading are: $E_a = 0.92$ eV, $E_b = 1.49$ eV, and $A = 1$. The optimal parameters for the periodically corrugated back reflector for $L_s = 200$ nm are: $L_g = 100$ nm, $\zeta = 0.50$, and $L_x = 500$ nm. The efficiency does not improve for $L_s > 600$ nm by the use of a periodically corrugated back reflector.

The highest efficiency predicted in Tables 5.2 and 5.3 is 17.07%, which arises when $L_s = 2200$ nm, $E_a = 0.91$ eV, $E_b = 1.49$ eV, and $A = 0.99$ for both flat ($L_g = 0$) and periodically corrugated back reflectors. The values of J_{sc}, V_{oc}, and FF corresponding to this optimal design are 36.72 mA cm^{-2}, 628 mV, and 74.0%, respectively. Relative to the optimal homogeneous CZTSSe layer (Section 4.2.5), the maximum efficiency increases from 11.84% to 17.07% (a

Table 5.2: Model-predicted parameters of the optimal CZTSSe solar cell with a specified value of $L_s \in [100, 2200]$ nm, when the CZTSSe layer is linearly graded according to Eq. (5.1) and the back reflector is flat ($L_g = 0$) [3]. The values of E_a and E_b provided pertain to the optical submodel.

L_s (nm)	E_a (eV)	E_b (eV)	A	J_{sc} (mA cm^{-2})	V_{oc} (mV)	FF (%)	η (%)
100	0.91	1.49	0.99	19.34	550	76.8	8.18
200	0.92	1.49	0.99	26.18	568	74.2	11.04
300	0.91	1.49	0.99	30.07	590	73.2	13.00
400	0.91	1.49	0.99	31.16	601	73.4	13.75
600	0.92	1.49	0.99	33.17	610	73.6	14.92
1200	0.93	1.49	0.99	35.02	617	73.5	15.90
2200	0.91	1.49	0.99	36.72	628	74.0	17.07

Table 5.3: Model-predicted parameters of the optimal CZTSSe solar cell with a specified value of $L_s \in [100, 2200]$ nm, when the CZTSSe layer is linearly graded according to Eq. (5.1) and the back reflector is periodically corrugated [3]. The values of E_a and E_b provided pertain to the optical submodel.

L_s (nm)	E_a (eV)	E_b (eV)	A	L_g (nm)	ζ	L_x (nm)	J_{sc} (mA cm^{-2})	V_{oc} (mV)	FF (%)	η (%)
100	0.92	1.49	0.99	100	0.50	510	20.24	544	76.0	8.44
200	0.92	1.49	1.00	100	0.50	500	27.42	572	74.5	11.69
300	0.91	1.49	0.99	100	0.50	510	29.88	592	74.0	13.01
400	0.92	1.49	0.99	100	0.51	550	31.39	603	73.0	13.91
600	0.91	1.49	1.00	100	0.50	502	32.98	612	73.6	14.87
1200	0.93	1.49	0.99	100	0.51	500	35.02	617	73.5	15.90
2200	0.91	1.49	0.99	100	0.51	500	36.72	628	74.0	17.07

relative increase of 44.1%) with forward grading of the CZTSSe layer; concurrently, J_{sc}, V_{oc}, as well as FF are also enhanced. *This result underscores the tremendous importance of optimal grading of the bandgap energy of the photon-absorbing layer.*

The optimal values of $E_a = 0.92 \pm 0.01$ eV and $A \simeq 1$ in Tables 5.2 and 5.3, and the optimal value of E_b is independent of L_s, whether the back reflector is flat or periodically corrugated. Also, the optimal corrugation parameters are very weakly dependent on L_s: $L_g = 100$ nm, $\zeta \simeq 0.505$, and $L_x \in [500, 550]$ nm.

Figure 5.4 presents the variations of E_g, χ, E_c, E_v, E_i, n, p, n_i, G, R, and N_f with z inside the optimal CZTSSe solar cell with a 2200-nm-thick forward-graded CZTSSe layer. Whether the back reflector is flat or periodically corrugated does not affect these spatial profiles. The dependences of E_c and E_i on z are almost linear, quite similar to the linear dependence of E_g on z, whereas n_i varies linearly with z such that it is small where E_g is large and vice versa. The electron-hole-pair generation rate G is higher near the front face $z = L_d - L_s$ and lower near the back face $z = L_d$ of the CZTSSe layer, which is in accord with the understanding [30, 31] that more charge carriers are generated in regions where E_g is lower, and vice versa, because less energy is required to excite a charge carrier from the valence band to the conduction band when E_g is lower. The J_{dev}–V_{ext} and P–V_{ext} characteristics of the solar cell yield $J_{dev} = 32.11$ mA cm^{-2} and $V_{ext} = 530$ mV for best performance along with FF $= 74\%$ and $\eta = 17.07\%$.

5.3.2   BACKWARD GRADING

Optoelectronic optimization yields $A = 0$ (i.e., a homogeneous bandgap) when Eq. (5.2) describes the grading profile [3]. Hence, Table 4.7 holds when the back reflector is flat and Table 4.8 when the back reflector is periodically corrugated. Thus, backward grading of the photon-absorbing layer of a CZTSSe solar cell is predicted by the coupled optoelectronic model [1] to be suboptimal, just like forward grading for CIGS solar cells.

5.4   ALGAAS SOLAR CELLS

Due to the quadratic dependence of the bandgap energy of AlGaAs on the compositional parameter ξ for $\xi \in [0.45, 1]$, it is convenient to specify the parameter space for optimization in terms of E_g rather than ξ for both the thin homogeneous p-AlGaAs layer and the thicker graded-bandgap n-AlGaAs photon-absorbing layer depicted in Fig. 5.5. With $L_m = 100$ nm, $L_{Pd} = 120$ nm, $L_{Ge} = 50$ nm, and $L_{Au} = 100$ nm in Fig. 4.7 fixed, the six-dimensional (i.e., $\tilde{N} = 6$) parameter space for maximizing the efficiency of an AlGaAs solar cell with a linearly bandgap-graded n-AlGaAs layer of a specified thickness $L_s \in [100, 2000]$ nm is: $\bar{\mathsf{E}}_g \in [1.424, 2.09]$ eV, $\mathsf{E}_a \in [1.424, 2.09]$ eV, $\mathsf{E}_b \in [1.424, 2.09]$ eV, $A \in [0, 1]$, $\zeta \in [0.05, 1]$, and $L_x \in [100, 1000]$ nm, where $\bar{\mathsf{E}}_g$ is the bandgap energy of p-AlGaAs. The common allowed range of $\bar{\mathsf{E}}_g$, E_a, and E_b is consistent with $\xi \in [0, 0.8]$, which is the ξ-range for which the permittivity ε of AlGaAs in the optical regime is available; see Fig. 4.9.

5.4.1   FORWARD GRADING

The bandgap energy in the photon-absorbing layer is smaller near its front face than near its back interface for $A > 0$, when Eq. (5.1) holds. Values of J_{sc}, V_{oc}, FF, and η for the optimal

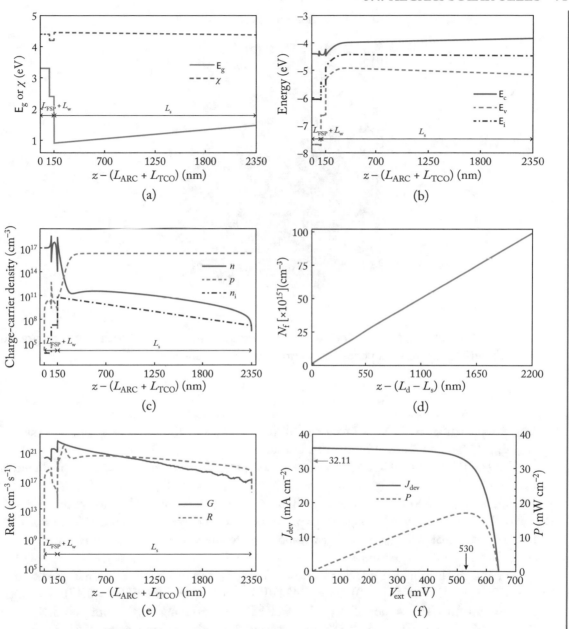

Figure 5.4: Variations of (a) E_g and χ; (b) E_c, E_v, and E_i; (c) n, p, and n_i; (d) N_f; and (e) G and R with z in the optimal CZTSSe solar cell with a 2200-nm-thick forward-graded CZTSSe photon-absorbing layer. (f) Plots of J_{dev} and P vs. V_{ext}. Values of J_{dev} and V_{ext} for maximum P are identified. The bandgap and geometric parameters are available in either Table 5.2 or Table 5.3; whether the back reflector is flat or periodically corrugated is inconsequential.

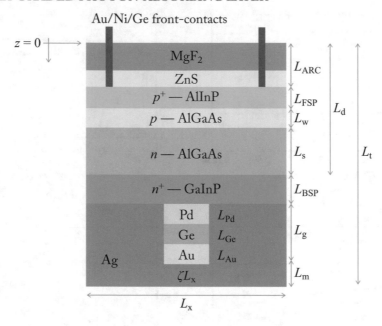

Figure 5.5: Schematic of the reference unit cell of the AlGaAs solar cell with a linearly graded photon-absorbing layer and a periodically corrugated back reflector.

designs are presented in Table 5.4 for seven different values of L_s. The corresponding values of $\bar{E}_g$, E_a, E_b, A, L_x, and ζ are also provided in that table.

For the thinnest n-AlGaAs layer ($L_s = 100$ nm), the maximum efficiency predicted is 21.0% with $\bar{E}_g = 1.424$ eV ($\bar{\xi} = 0$), $E_a = 1.424$ eV ($\xi_{min} = 0$), $E_b = 1.98$ eV ($\xi_{max} = 0.45$), $A = 1.0$, and $L_x = 500$ nm. A relative enhancement of 13.5% over the maximum efficiency 18.5% in Table 4.11 for the homogeneous photon-absorbing layer of the same thickness is predicted. The values of J_{sc}, V_{oc}, and FF corresponding to the optimal design are 16.8 mA cm^{-2}, 1399 mV, and 89.3%, respectively.

For the thickest n-AlGaAs layer ($L_s = 2000$ nm), the maximum efficiency predicted is 33.1% with $\bar{E}_g = 1.424$ eV ($\bar{\xi} = 0$), $E_a = 1.424$ eV ($\xi = 0$), $E_b = 1.98$ eV ($\xi = 0.45$), $A = 0.99$, and $L_x = 500$ nm. The values of J_{sc}, V_{oc}, and FF corresponding to this optimal design are 24.7 mA cm^{-2}, 1507 mV, and 88.8%, respectively. A relative enhancement of 14.9% is predicted with linear bandgap-grading of the n-AlGaAs absorber layer over the optimal efficiency of 28.8% with the homogeneous n-AlGaAs layer in Table 4.9. For this optimal design, the p-AlGaAs layer is really a p-GaAs layer, but the n-AlGaAs absorber layer is different from a n-GaAs absorber layer. Although V_{oc} is significantly higher with the linearly graded bandgap compared to the homogeneous bandgap, J_{sc} is lower with the linearly graded bandgap.

Table 5.4: Model-predicted parameters of the optimal AlGaAs solar cell with a specified value of $L_s \in [100, 2000]$ nm, when the n-AlGaAs absorber layer is linearly graded according to Eq. (5.1)

L_s (nm)	$\bar{E}_g$ (eV)	E_a (eV)	E_b (eV)	A	ζ	L_x (nm)	J_{sc} (mA cm^{-2})	V_{oc} (mV)	FF (%)	η (%)
100	1.424	1.424	1.98	0.99	0.05	500	16.8	1399	89.3	21.0
200	1.424	1.424	1.98	0.99	0.05	510	19.0	1422	81.9	22.2
300	1.424	1.424	1.98	1.00	0.05	502	19.1	1441	85.3	23.5
400	1.424	1.424	1.98	0.99	0.05	510	19.8	1453	86.5	24.9
500	1.424	1.424	1.98	0.98	0.05	500	20.5	1462	87.1	26.1
1000	1.424	1.424	1.98	0.99	0.05	500	22.7	1486	88.3	29.8
2000	1.424	1.424	1.98	1.00	0.05	500	24.7	1507	88.8	33.1

Similar to the data for the homogeneous photon-absorbing layer provided in Table 4.11, the optimal designs in Table 5.4 have $A = 0.99 \pm 0.1$, $L_x = 505 \pm 5$ nm, $\zeta = 0.05$, regardless of the value of L_s. Also, both E_a and $\bar{E}_g$ are independent of L_s in both tables. Also, for both homogeneous and linearly graded n-AlGaAs layers, E_a is at its minimum allowed value. However, the value of $\bar{E}_g$ is at its minimum allowed value when the photon-absorbing layer is linearly graded, but at its maximum allowed value when the photon-absorbing layer is homogeneous. The value of $E_b = 1.98$ eV is independent of L_s in Table 4.11, the latter being significantly lower than its maximum allowed value. Most importantly, just as for CIGS and CZTSSe solar cells, *the optimal grading of the bandgap energy of the photon-absorbing layer will improve the efficiency of the AlGaAs solar cell.*

The highest efficiency of 33.1% for the solar cell whose n-AlGaAs photon-absorbing layer has forward-graded bandgap energy is $L_s = 2000$ nm. Spatial profiles of E_g, χ, E_c, E_v, E_i, n, p, n_i, G, R, and N_f are available in Fig. 5.6. The dependences of E_c and E_i on z are quite similar to that of E_g in the n-AlGaAs photon-absorbing layer, whereas n_i varies linearly with z such that it is small where E_g is large and vice versa. The electron-hole-pair generation rate G is higher near the front face $z = L_d - L_s$ and lower near the back face $z = L_d$ of the n-AlGaAs layer, which is as expected [30]. The $J_{dev}-V_{ext}$ and $P-V_{ext}$ characteristics of the solar cell yield $J_{dev} = 23.9$ mA cm^{-2} and $V_{ext} = 1375$ mV for best performance along with FF = 88.8% and $\eta = 33.1\%$.

5.4.2   BACKWARD GRADING

The coupled optoelectronic model did not perform satisfactorily for backward grading of the bandgap energy of the n-AlGaAs layer [4].

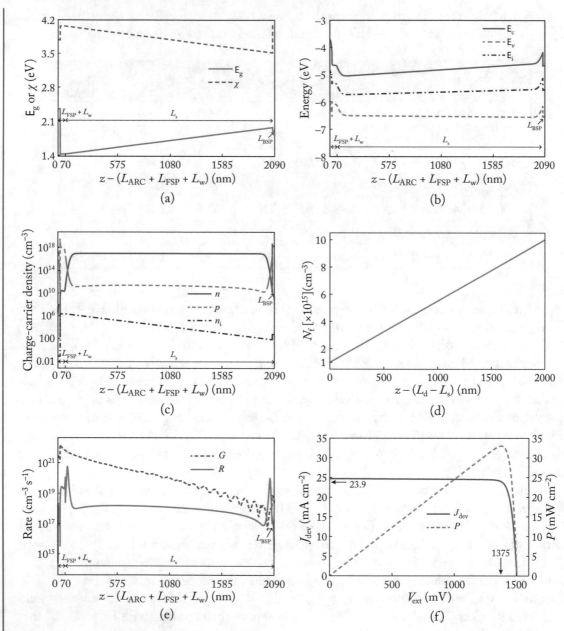

Figure 5.6: Variations of (a) E_g and χ; (b) E_c, E_v, and E_i; (c) n, p, and n_i; (d) N_f; and (e) G and R with z in the optimal AlGaAs solar cell with a 2000-nm-thick forward-graded n-AlGaAs photon-absorbing layer. (f) Plots of J_{dev} and P vs. V_{ext}. Values of J_{dev} and V_{ext} for maximum P are identified. The bandgap and geometric parameters are available in Table 5.4.

5.5 BIBLIOGRAPHY

[1] T. H. Anderson, B. J. Civiletti, P. B. Monk, and A. Lakhtakia, Coupled optoelectronic simulation and optimization of thin-film photovoltaic solar cells, *Journal of Computational Physics*, 407:109242 (2020). 83, 84, 90

T. H. Anderson, B. J. Civiletti, P. B. Monk, and A. Lakhtakia, Coupled optoelectronic simulation and optimization of thin-film photovoltaic solar cells, *Journal of Computational Physics*, 418:109561 (2020) (corrigendum).

[2] F. Ahmad, T. H. Anderson, P. B. Monk, and A. Lakhtakia, Efficiency enhancement of ultrathin CIGS solar cells by optimal bandgap grading, *Applied Optics*, 58:6067–6078 (2019). 83, 84, 87

F. Ahmad, T. H. Anderson, P. B. Monk, and A. Lakhtakia, Efficiency enhancement of ultrathin CIGS solar cells by optimal bandgap grading, *Applied Optics*, 59:2615 (2020) (erratum).

[3] F. Ahmad, A. Lakhtakia, T. H. Anderson, and P. B. Monk, Towards highly efficient thin-film solar cells with a graded-bandgap CZTSSe layer, *Journal of Physics: Energy*, 2:025004 (2020). 83, 84, 88, 89, 90

F. Ahmad, A. Lakhtakia, T. H. Anderson, and P. B. Monk, Towards highly efficient thin-film solar cells with a graded-bandgap CZTSSe layer, *Journal of Physics: Energy*, 2:039501 (2020) (corrigendum).

[4] F. Ahmad, A. Lakhtakia, and P. B. Monk, Optoelectronic optimization of graded-bandgap thin-film AlGaAs solar cells, *Applied Optics*, 59:1018–1027 (2020). 83, 84, 93

[5] L. M. Anderson, Harnessing surface plasmons for solar energy conversion, *Proc. of SPIE*, 408:172–178 (1983). 83

[6] H. A. Atwater and A. Polman, Plasmonics for improved photovoltaic devices, *Nature Materials*, 9:205–213 (2010). 83

[7] A. S. Hall, M. Faryad, G. D. Barber, L. Liu, S. Erten, T. S. Mayer, A. Lakhtakia, and T. E. Mallouk, Broadband light absorption with multiple surface plasmon polariton waves excited at the interface of a metallic grating and photonic crystal, *ACS Nano*, 7:4995–5007 (2013). 83

[8] F.-J. Haug, K. Söderström, A. Naqavi, and C. Ballif, Excitation of guided-mode resonances in thin film silicon solar cells, *Materials Research Society Proceedings*, 1321:123–128 (2011). 83

[9] T. Khaleque and R. Magnusson, Light management through guided-mode resonances in thin-film silicon solar cells, *Journal of Nanophotonics*, 8:083995 (2014). 83

[10] F. Ahmad, T. H. Anderson, B. J. Civiletti, P. B. Monk, and A. Lakhtakia, On optical absorption peaks in a nonhomogeneous thin-film solar cell with a two-dimensional periodically corrugated metallic backreflector, *Journal of Nanophotonics*, 12:016017 (2018). 83

[11] F. Ahmad, T. H. Anderson, P. B. Monk, and A. Lakhtakia, Optimization of light trapping in ultrathin nonhomogeneous $CuIn_{1-\xi}Ga_{\xi}Se_2$ solar cell backed by 1D periodically corrugated backreflector, *Proc. of SPIE*, 10731:107310L (2018). 83

[12] B. J. Civiletti, T. H. Anderson, F. Ahmad, P. B. Monk, and A. Lakhtakia, Optimization approach for optical absorption in three-dimensional structures including solar cells, *Optical Engineering*, 57:057101 (2018). 83

[13] T. H. Anderson, A. Lakhtakia, and P. B. Monk, Optimization of nonhomogeneous indium-gallium-nitride Schottky-barrier thin-film solar cells, *Journal of Photonics for Energy*, 8:034501 (2018). 83

[14] M. Gloeckler and J. R. Sites, Potential of submicrometer thickness $Cu(In,Ga)Se_2$ solar cells, *Journal of Applied Physics*, 98:103703 (2005). 83

[15] K. Woo, Y. Kim, W. Yang, K. Kim, I. Kim, Y. Oh, J. K. Kim, and J. Moon, Bandgap-graded $Cu_2ZnSn(S_{1-x},Se_x)_4$ solar sells fabricated by an ethanol-based, particulate precursor ink route, *Scientific Reports*, 3:03069 (2013). 83

[16] I. M. Dharmadasa, A. A. Ojo, H. I. Salim, and R. Dharmadasa, Next generation solar cells based on graded bandgap device structures utilising rod-type nano-materials, *Energies*, 8:5440–5458 (2015). 83

[17] K.-J. Yang, D.-H. Son, S.-J. Sung, J.-H. Sim, Y.-I Kim, S.-N. Park, D.-H. Jeon, J. Kim, D.-K. Hwang, C. W. Jeon, D. Nam, H. Cheong, J.-K. Kang, and D.-H. Kim, A bandgap-graded CZTSSe solar cell with 12.3% efficiency, *Journal of Materials Chemistry A*, 4:10151–10158 (2016). 83

[18] D.-K. Hwang, B.-S. Ko, D.-H. Jeon, J.-K. Kang, S.-J. Sung, K.-J. Yang, D. Nam, S. Cho, H. Cheong, and D.-H. Kim, Single-step sulfo-selenization method for achieving low open circuit voltage deficit with band gap front-graded Cu_2ZnSnS,Se_4 thin films, *Solar Energy Materials and Solar Cells*, 161:162–169 (2017). 83

[19] J. A. Hutchby, High-efficiency graded band-gap $Al_xGa_{1-x}As$–GaAs solar cell, *Applied Physics Letters*, 26:457–459 (1975). 83

[20] J. Song, S. S. Li, C. H. Huang, O. D. Crisalle, and T. J. Anderson, Device modeling and simulation of the performance of Cu(In$_{1-x}$,Ga$_x$)Se$_2$ solar cells, *Solid-State Electronics*, 48:73–79 (2004). 83

[21] S. H. Song, K. Nagaich, E. S. Aydil, R. Feist, R. Haley, and S. A. Campbell, Structure optimization for a high efficiency CIGS solar cell, *Proc. of 35th IEEE Photovoltaic Specialists Conference (PVSC)*, pages 2488–2492, Honolulu, HI, June 20–25, 2010. 83

[22] D. Hironiwa, M. Murata, N. Ashida, Z. Tang, and T. Minemoto, Simulation of optimum band-gap grading profile of Cu$_2$ZnSn(S,Se)$_4$ solar cells with different optical and defect properties, *Japanese Journal of Applied Physics*, 53:071201 (2014). 83

[23] O. K. Simya, A. Mahaboobbatcha, and K. Balachander, Compositional grading of CZTSSe alloy using exponential and uniform grading laws in SCAPS-ID simulation, *Superlattices and Microstructures*, 92:285–293 (2016). 83

[24] M. Chadel, A. Chadel, M. M. Bouzaki, M. Aillerie, B. Benyoucef, and J.-P. Charles, Optimization by simulation of the nature of the buffer, the gap profile of the absorber and the thickness of the various layers in CZTSSe solar cells, *Materials Research Express*, 4:115503 (2017). 83

[25] S. Mohammadnejad and A. B. Parashkouh, CZTSSe solar cell efficiency improvement using a new band-gap grading model in absorber layer, *Applied Physics A*, 123:758 (2017). 83

[26] M. Gloeckler and J. R. Sites, Band-gap grading in Cu(In,Ga)Se$_2$ solar cells, *Journal of Physics and Chemistry of Solids*, 66:1891–1894 (2005). 83

[27] H. Ferhati and F. Djeffal, Graded band-gap engineering for increased efficiency in CZTS solar cells, *Optical Materials*, 76:393–399 (2018).

[28] R. Storn and K. Price, Differential evolution—a simple and efficient heuristic for global optimization over continuous spaces, *Journal of Global Optimization*, 11:341–359 (1997). 84

[29] F. Ahmad, Optoelectronic modeling and optimization of graded-bandgap thin-film solar cells, Ph.D. Dissertation (The Pennsylvania State University, University Park, PA, 2020). 84

[30] S. J. Fonash, *Solar Cell Device Physics*, 2nd ed. (Academic Press, Burlington, MA, 2010). 90, 93

[31] I. L. Repins, L. Mansfield, A. Kanevce, S. A. Jensen, D. Kuciauskas, S. Glynn, T. Barnes, W. Metzger, J. Burst, C.-S. Jiang, P. Dippo, S. Harvey, G. Teeter, C. Perkins, B. Egaas, A.

Zakutayev, J.-H. Alsmeier, T. Lußky, L. Korte, R. G. Wilks, M. Bär, Y. Yan, S. Lany, P. Zawadzki, J.-S. Park, and S. Wei, Wild band edges: The role of bandgap grading and band-edge fluctuations in high-efficiency chalcogenide devices, *Proc. of 43rd IEEE Photovoltaics Specialists Conference (PVSC)*, pages 309–314, Portland, OR, June 5–10, 2016. 90

CHAPTER 6

Nonlinearly Graded Photon-Absorbing Layer

The bandgap energy of a compound semiconductor depends on its composition, as exemplified by the compositional parameter ξ which quantifies the gallium-indium ratio in $CuIn_{1-\xi}Ga_\xi Se_2$, the sulfur-selenium ratio in $Cu_2ZnSn(S_\xi Se_{1-\xi})_4$, and the aluminum-gallium ratio in $Al_\xi Ga_{1-\xi}As$. Therefore, bandgap grading of the photon-absorbing layer can be accomplished via compositional grading when the photon-absorbing material is a compound semiconductor.

The coupled optoelectronic model of Chapters 2 and 3 [1] has been applied in Chapter 5 to determine the efficiency of a thin-film solar cell with a photon-absorbing layer whose bandgap energy E_g is linearly graded in the thickness direction. For solar cells of three different types, the clear conclusion is that optimal linear grading of the bandgap energy of the photon-absorbing layer will improve the power-conversion efficiency.

6.1 NONLINEAR GRADING PROFILES

More complicated grading profiles can be devised for even better performance than linear grading profiles, as exemplified by the family of nonlinear grading profiles [2–4]

$$E_g(z) = E_a + A\left(E_b - E_a\right)\left[\frac{1}{2}\left(\sin\left\{2\pi\left[K\frac{z-(L_d-L_s)}{L_s} - \nu\right]\right\} + 1\right)\right]^\alpha,$$

$$z \in (L_d - L_s, L_d),$$

$$(6.1)$$

where $\nu \in [0, 1)$ quantifies a relative phase shift, $\alpha > 0$ is a shaping parameter, and $K \geq 0$ is a cycle number that need not be an integer. Four examples are shown in Fig. 6.1. With $\nu = 0$, $\alpha = 1$, and $K \ll 1$, Eq. (6.1) is approximated by the linear forward profile of Eq. (5.1).

The nonlinear bandgap profile of the photon-absorbing layer and the dimensions of the back reflector have been optimized for CIGS [2], CZTSSe [3], and AlGaAs [4] solar cells using the coupled optoelectronic model and the differential evolution algorithm [5, 6], as discussed next in this chapter. All necessary optical and electrical parameters are provided in Chapter 4.

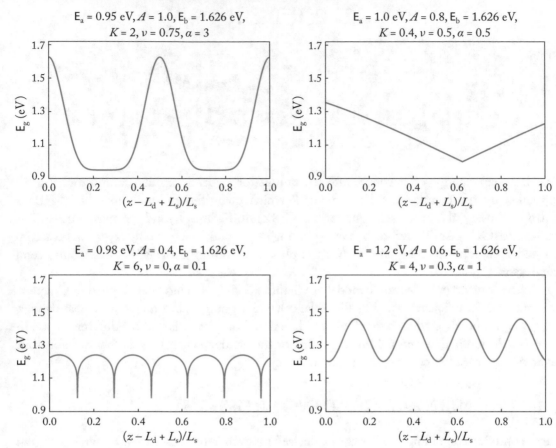

Figure 6.1: Four representative examples of the nonlinearly graded bandgap of a CIGS photon-absorbing layer.

6.2   CIGS SOLAR CELLS

Figure 6.2 is a schematic of the CIGS solar cell. With $E_b = 1.626$ eV in Eq. (6.1) fixed, the eight-dimensional (i.e., $\tilde{N} = 8$) parameter space for maximizing the efficiency of a CIGS solar cell with the nonlinearly bandgap-graded CIGS layer of a specified thickness $L_s \in [100, 2200]$ nm is as follows: $E_a \in [0.947, 1.626]$ eV, $A \in [0, 1]$, $\alpha \in [0, 8]$, $K \in [0, 8]$, $\nu \in [0, 1]$, $L_g \in [1, 550]$ nm, $\zeta \in (0, 1)$, and $L_x \in [100, 1000]$ nm. Optimization must be carried out subject to the constraint $E_b \geq E_a$. The parameter space is six-dimensional when the back reflector is flat, because L_g and ζ then become irrelevant.

Table 6.1 presents η optimized for nine representative values of L_s, along with the corresponding values of E_a, A, α, K, ν, L_g, ζ, L_x, J_{sc}, V_{oc}, and FF. For $L_s = 100$ nm, the maximum

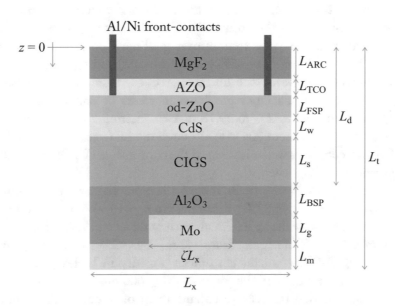

Figure 6.2: Schematic of the reference unit cell of the CIGS solar cell with a nonlinearly graded photon-absorbing layer and a periodically corrugated back reflector. The corrugation height $L_g = 0$ for a flat back reflector.

Table 6.1: Model-predicted parameters of the optimal CIGS solar cell with a specified value of $L_s \in [100, 2200]$ nm, when the CIGS layer is nonlinearly graded according to Eq. (6.1) with $E_b = 1.626$ eV and the back reflector is periodically corrugated

L_s (nm)	E_a (eV)	A	α	K	ν	L_g (nm)	ζ	L_x (nm)	J_{sc} (mA cm^{-2})	V_{oc} (mV)	FF (%)	η (%)
100	0.96	0.990	6.14	0.75	0.75	101	0.50	510	17.04	969	71	12.37
200	0.95	1.000	6.00	1.50	0.75	101	0.48	500	24.12	1007	69	16.89
300	0.95	0.980	6.00	1.50	0.74	102	0.49	502	25.98	1023	71	19.01
400	0.95	0.980	6.00	1.50	0.75	111	0.50	500	27.17	1033	73	20.66
500	0.95	0.990	6.00	1.50	0.76	111	0.50	520	28.23	1040	74	21.90
600	0.95	0.992	6.00	1.50	0.75	106	0.48	510	29.18	1045	75	22.89
900	0.95	0.992	6.00	1.50	0.75	106	0.48	510	30.86	1057	76	24.98
1200	0.95	0.980	6.00	1.50	0.75	106	0.48	510	32.02	1063	77	26.33
2200	0.95	0.980	6.00	1.50	0.75	106	0.48	510	33.16	1070	78	27.70

efficiency predicted is 12.37%, a massive relative increase of 70.62% over the maximum efficiency of 7.25% in Table 4.4 for the solar cell with a homogeneous photon-absorbing layer. This huge enhancement must be due to nonlinear grading. Concurrently, J_{sc} increases from 14.89 mA cm^{-2} to 17.04 mA cm^{-2} (14.43% relative increase) and V_{oc} from 624 mV to 969 mV (55.28% relative increase); however, the fill factor reduces from 78% to 71%. Likewise, for $L_s = 600$ nm, the maximum efficiency predicted increases by 65.98% to $\eta = 22.89\%$ from $\eta = 13.79\%$, J_{sc} increases by 18.32% to 29.18 mA cm^{-2} from 24.66 mA cm^{-2}, and V_{oc} increases by 48.43% to 1045 mV from 704 mV, although the fill factor reduces to 75% from 79%. The 22.89% efficiency compares well with that of the conventional CIGS solar cell with a 2200-nm-thick homogeneous CIGS layer [7].

The highest maximum $\eta = 27.7\%$ in Table 6.1 is predicted for the 2200-nm-thick CIGS layer, a relative enhancement of 46.32% with respect to $\eta = 18.93\%$ predicted for the homogeneous CIGS layer in Table 4.4. The short-circuit current density increases from 31.11 mA cm^{-2} by 6.5% to 33.16 mA cm^{-2}, V_{oc} increases from 742 mV by 44.2% to 1070 mV, but the fill factor reduces to 78% from 82%. The predicted efficiencies provided in Table 6.1 reduce slightly when the back reflector is flat (i.e., $L_g = 0$), but only for $L_s \leq 600$ nm.

The overall trend is that the relative enhancement of η, due to the nonlinearly graded bandgap, decreases with the increase of L_s, on optoelectronic optimization. Notably, V_{oc} increases significantly without J_{sc} decreasing, as has been previously observed for solar cells with a linearly graded CIGS layer [8, 9]. Clearly, the grading was suboptimal in those studies.

The relative enhancement in V_{oc} is almost the same for linear grading (Table 5.1) as for nonlinear grading (Table 6.1) of the CIGS layer in comparison to the homogenous CIGS layer (Tables 4.4). However, whereas nonlinear grading enhances J_{sc}, linear grading depresses J_{sc}, in comparison to the homogeneous CIGS layer. As the reduction of J_{sc} does not entirely overcome the enhancement of V_{oc} for linear grading, the efficiencies in Table 5.1 exceed their counterparts in Tables 4.4; of course, the efficiencies in Table 6.1 are even higher. Thus, *nonlinear grading of the bandgap energy of the photon-absorbing layer is considerably more efficient than both no grading and linear grading.*

Optimal values of A range from 0.98 to 1 in Table 6.1, which is in contrast to $A < 0.04$ delivered by optical optimization [10]. The largest value of A provides the largest possible bandgap variation in the graded CIGS layer. Thus, optoelectronic optimization promotes nonlinear nonhomogeneity of the bandgap energy with a large amplitude, whereas optical optimization severely suppresses nonhomogeneity of the bandgap energy. The inescapable conclusions are that

- optical optimization is seriously deficient and

- optoelectronic optimization is essential for graded-bandgap solar cells.

The optimal designs in Table 6.1 do not vary much with L_s. Thus, $E_a \simeq 0.955$ eV, $A = 0.99 \pm 0.1$, $\alpha \simeq 6.05$, $\nu \simeq 0.75$, $\zeta = 0.49 \pm 0.01$, and $L_x \simeq 505$ nm. There is somewhat greater spread in the values of K, L_g, and L_x. The maximum efficiency therefore is mostly dependent on L_s and increases as the photon-absorbing layer is made thicker.

The highest efficiency in Table 6.1 is 27.7%, predicted by the coupled optoelectronic model for the CIGS solar cell with a nonlinearly graded 2200-nm-thick CIGS layer. Figure 6.3 presents the variations of E_g, χ, E_c, E_v, E_i, n, p, n_i, G, R, and N_f with z. Whether the back reflector is flat or periodically corrugated does not affect these spatial profiles.

The magnitude of $E_g(z)$ in Fig. 6.3 is large in the proximity of the plane $z = L_d - L_s$, which elevates V_{oc}. The regions in which the bandgap profile is flat and low ($\sim$1 eV) are of substantial thickness, these regions being responsible for elevating the electron-hole-pair generation rate because less energy is required to excite a charge carrier to cross the bandgap when E_g is lower [11]. The spatial profiles of G and R confirm that the net production of charge carriers is boosted in regions of the uniformly low E_g. The nonlinear grading close to the back face $z = L_d$ adds an additional drift electric field to reduce the recombination rate, thereby supplementing the role of the Al_2O_3 passivation layer. Hence, the efficiency is significantly higher for nonlinear grading (22.89%) than for linear grading (15.79%) of the bandgap energy in the CIGS layer.

The spatial profiles of E_c and E_i in Fig. 6.3 are similar to that of E_g and provide the conditions to enhance the generation rate G [8]. The intrinsic carrier density varies such that n_i is small where E_g is large and vice versa. Furthermore, the spatial profile of the recombination rate R follows that of the defect density N_f. The J_{dev}–V_{ext} and P–V_{ext} characteristics of the solar cell yield $J_{dev} = 24.72$ mA cm^{-2} and $V_{ext} = 926$ mV for best performance along with FF = 78%.

6.3   CZTSSe SOLAR CELLS

A schematic of the CZTSSe solar cell is provided in Fig. 6.4. With $E_b = 1.49$ eV in Eq. (6.1) fixed, the eight-dimensional (i.e., $\tilde{N} = 8$) parameter space for maximizing the efficiency of a CZTSSe solar cell with the nonlinearly bandgap-graded CZTSSe layer of a specified thickness $L_s \in [100, 2200]$ nm is as follows: $E_a \in [0.91, 1.49]$ eV, $A \in [0, 1]$, $\alpha \in [0, 8]$, $K \in [0, 8]$, $v \in [0, 1]$, $L_g \in [1, 550]$ nm, $\zeta \in (0, 1)$, and $L_x \in [100, 1000]$ nm. The constraint $E_b \geq E_a$ must be applied during optimization. The parameter space is six-dimensional when the back reflector is flat.

The values of J_{sc}, V_{oc}, FF, and η predicted by the coupled optoelectronic model are presented in Table 6.2 for eight representative values of L_s. The values of E_a, A, K, α, v, L_g, ζ, and L_x for the optimal designs are also provided in the same table. Periodic corrugation of the back reflector slightly improves the efficiency if the photon-absorbing layer is ultrathin. For $L_s = 200$ nm, the optimal efficiency predicted is 17.48% with a flat back reflector [3] and 17.83% with a periodically corrugated back reflector. However, the optimal efficiency predicted is 19.56% for $L_s = 2200$ nm, regardless of the geometry of the back reflector. The effect of periodic corrugation remains the same as for the cases of the homogeneous bandgap in Section 4.2 and the linearly graded bandgap in Section 5.3: very small improvement for thin CZTSSe layers and no improvement beyond $L_s \simeq 600$ nm.

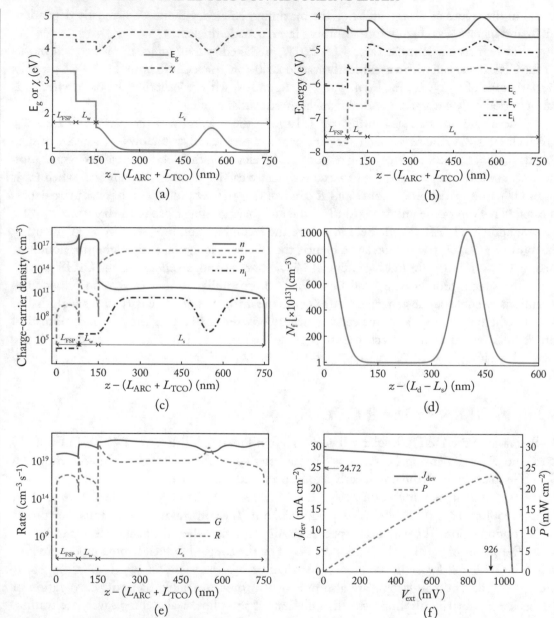

Figure 6.3: Variations of (a) E_g and χ; (b) E_c, E_v, and E_i; (c) n, p, and n_i; (d) N_f; and (e) G and R with z in the optimal CIGS solar cell with a 2200-nm-thick nonlinearly graded CIGS photon-absorbing layer. (f) Plots of J_{dev} and P vs. V_{ext}. Values of J_{dev} and V_{ext} for maximum P are identified. The bandgap and geometric parameters are available in Table 6.1; whether the back reflector is flat or periodically corrugated is inconsequential.

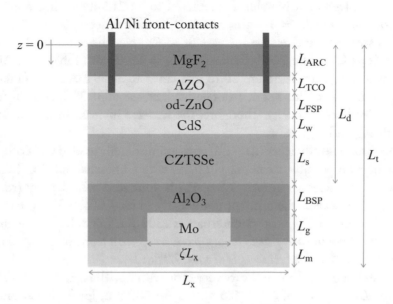

Figure 6.4: Schematic of the reference unit cell of the CZTSSe solar cell with a nonlinearly graded photon-absorbing layer and a periodically corrugated back reflector. The corrugation height $L_g = 0$ for a flat back reflector.

Table 6.2: Model-predicted parameters of the optimal CZTSSe solar cell with a specified value of $L_s \in [100, 2200]$ nm, when the CZTSSe layer is nonlinearly graded according to Eq. (6.1) with $E_b = 1.49$ eV and the back reflector is periodically corrugated. The values of E_a provided pertain to the optical submodel.

L_s (nm)	E_a (eV)	A	α	K	v	L_g (nm)	ζ	L_x (nm)	J_{sc} (mA cm^{-2})	V_{oc} (mV)	FF (%)	η (%)
100	0.92	0.98	6	3	0.75	100	0.50	500	25.72	701	78.7	14.22
200	0.92	0.99	6	3	0.75	100	0.51	510	32.99	716	77.5	17.83
300	0.92	0.98	6	2	0.75	100	0.51	510	35.15	745	74.7	19.58
400	0.92	0.98	6	2	0.75	100	0.51	510	36.32	762	74.4	20.62
600	0.92	0.98	6	2	0.75	100	0.50	500	37.23	771	74.8	21.47
870	0.92	0.98	6	2	0.75	100	0.50	500	37.39	772	75.2	21.74
1200	0.92	0.98	6	2	0.75	100	0.51	510	37.08	766	74.8	21.26
2200	0.92	0.98	6	2	0.75	100	0.51	510	36.45	736	72.8	19.56

The highest efficiency achievable is predicted to be 21.74% with a nonlinearly graded CZTSSe layer of thickness $L_s = 870$ nm, whether the back reflector is flat or periodically corrugated [3]. This amounts to a relative increase of 83.6% over the optimal efficiency of 11.84% with a homogeneous CZTSSe layer of thickness $L_s = 1200$ nm (Table 4.8). Along with the increase in efficiency, J_{sc} increases from 30.13 mA cm^{-2} to 37.39 mA cm^{-2} (a relative increase of 24.0%), V_{oc} from 558 mV to 772 mV (a relative increase of 38.3%), and FF from 70.3% to 75.2% (a relative increase of 6.9%). Therefore, the massive increase in η is due to substantial increases in both V_{oc} and J_{sc}.

The highest possible efficiency (21.74%) with a sinusoidally graded CZTSSe layer is 27.3% higher than the highest possible efficiency (17.07%) with a linearly graded CZTSSe layer (Table 5.3). Whereas J_{sc} increases by only 1.8%, V_{oc} is enhanced considerably from 628 mV by 22.9% to 772 mV. However, the optimal nonlinearly graded CZTSSe layer is only 870-nm thick, but its optimal linearly graded counterpart is 2200-nm thick. Just as for the CIGS solar cell in Section 6.2, *the nonlinearly graded bandgap is more efficient than the homogeneous and linearly graded bandgaps for all considered thicknesses of the CZTSSe layer.*

The optimal designs in Table 6.2 vary very little with L_s. Thus, $E_a \simeq 0.905$ eV, $A \simeq 0.985$, $\alpha = 6$, $\nu = 0.75$, $L_g = 100$ nm, $\zeta \simeq 0.505$ and $L_x \simeq 505$ nm, for both flat and periodically corrugated back reflectors. However, $K = 3$ for $L_s \in \{100, 200\}$ nm but $K = 2$ for $L_s \geq 300$ nm.

Figure 6.5 presents the variations of E_g, χ, E_c, E_v, E_i, n, p, n_i, G, R, and N_f with z inside the most efficient CZTSSe solar cell ($L_s = 870$ nm). With $E_a = 0.92$ eV and $A = 0.98$, $E_g(z) \in [0.92, 1.486]$ eV. The magnitude of $E_g(z)$ is large near both faces of the CZTSSe layer, which elevates V_{oc} [12]. Furthermore, bandgap grading in the proximity of the back face of the CZTSSe layer keeps the minority carriers away from that face to reduce the recombination rate [13] and improve carrier collection due to the drift field provided by the bandgap grading [14]. The regions in which E_g is small are of substantial thickness, and it is those very regions that are responsible for increasing the electron-hole-pair generation rate [11, 15], because less energy is required to excite an electron-hole-pair across a narrower bandgap. Indeed, G is higher in regions with lower E_g and vice versa. Thus, this bandgap profile is ideal for the enhancement of V_{oc} while maintaining a large J_{sc}.

The spatial profiles of E_c and E_i are similar to that of E_g in Fig. 6.5, and n_i varies such that it is large where E_g is small and vice versa. The higher recombination rate in the 60-nm-thick middle region of the CZTSSe layer is due to higher defect density N_f caused by higher sulfur content. The coupled optoelectronic model predicts best performance when $V_{ext} = 659$ mV and $J_{dev} = 32.72$ mA cm^{-2}.

6.4   ALGAAS SOLAR CELLS

Figure 6.6 is a schematic of an AlGaAs solar cell. The nine-dimensional (i.e., $\tilde{N} = 9$) parameter space for maximizing the efficiency of an AlGaAs solar cell with the nonlinearly bandgap-graded AlGaAs layer of a specified thickness $L_s \in [100, 2200]$ nm is as follows: $\bar{E}_g \in$

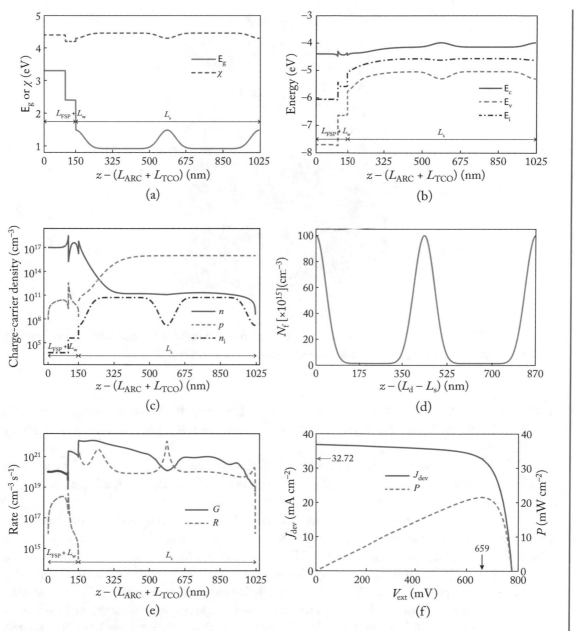

Figure 6.5: Variations of (a) E_g and χ; (b) E_c, E_v, and E_i; (c) n, p, and n_i; (d) N_f; and (e) G and R with z in the optimal CZTSSe solar cell with a 870-nm-thick nonlinearly graded CZTSSe photon-absorbing layer. (f) Plots of J_{dev} and P vs. V_{ext}. Values of J_{dev} and V_{ext} for maximum P are identified. The bandgap and geometric parameters are available in Table 6.2; whether the back reflector is flat or periodically corrugated is inconsequential.

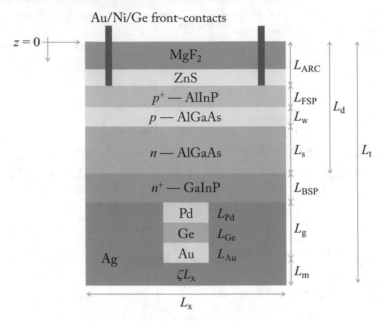

Figure 6.6: Schematic of the reference unit cell of the AlGaAs solar cell with a nonlinearly graded photon-absorbing layer and localized ohmic back contacts.

[1.424, 2.09] eV, $E_a \in [1.424, 2.09]$ eV, $E_b \in [1.424, 2.09]$ eV, $A \in [0, 1]$, $\alpha \in [0, 8]$, $K \in [0, 8]$, $\nu \in [0, 1]$, $\zeta \in [0.05, 1]$, and $L_x \in [100, 1000]$ nm. The constraint $E_b \geq E_a$ must be applied during optimization.

Maximization of η with $L_{Ag} = 100$ nm fixed yields values of J_{sc}, V_{oc}, FF, and η presented in Table 6.3 for seven different values of L_s [4]. The values of $\bar{E}_g$, E_a, E_b, A, α, K, ν, L_x, and ζ for the optimal designs are also provided in the same table.

For the thinnest n-AlGaAs photon-absorbing layer ($L_s = 100$ nm), the maximum efficiency predicted by the coupled optoelectronic model is 21.2%. This amounts to a relative enhancement of 14.5% over the maximum efficiency of 18.5% for the homogeneous n-AlGaAs layer (Table 4.11). For the thickest n-AlGaAs photon-absorbing layer ($L_s = 2000$ nm), the maximum efficiency predicted is 34.5%. Thus, a relative enhancement of 19.8% is possible with nonlinear bandgap grading of the n-AlGaAs layer over the maximum efficiency of 28.8% with the homogeneous n-AlGaAs layer (Table 4.11), and a relative enhancement of 4.2% over the maximum efficiency of 33.1% with the linearly graded n-AlGaAs layer (Table 5.4).

Just as in Section 5.4.1 for the linearly graded bandgap, although V_{oc} is significantly higher with the nonlinearly graded bandgap compared to the homogeneous bandgap, J_{sc} is lower with the nonlinearly graded bandgap. Whereas J_{sc} is almost the same for both linearly and nonlinearly graded n-AlGaAs layers, V_{oc} is notably higher with the nonlinearly graded n-AlGaAs

Table 6.3: Model-predicted parameters of the optimal AlGaAs solar cell with a specified value of $L_s \in [100, 2000]$ nm, when the n-AlGaAs layer is nonlinearly graded according to Eq. (6.1) and $L_{Ag} = 100$ nm

L_s (nm)	$\bar{E}_g$ (eV)	E_a (eV)	E_b (eV)	A	α	K	ν	L_x (nm)	ζ	J_{sc} (mA cm^{-2})	V_{oc} (mV)	FF (%)	η (%)
100	2.09	1.424	1.98	1.00	6	3	0.75	510	0.05	16.1	1455	90.3	21.2
200	2.09	1.424	1.98	1.00	6	1	0.75	520	0.05	19.2	1471	80.2	22.6
300	2.06	1.424	1.98	1.00	6	1	0.74	512	0.05	19.7	1486	80.2	23.5
400	2.09	1.424	1.98	1.00	6	1	0.75	509	0.05	20.2	1497	82.0	24.8
500	2.08	1.424	1.98	1.00	6	1	0.75	524	0.05	20.8	1505	83.0	26.0
1000	2.09	1.424	1.98	1.00	6	2	0.75	516	0.05	22.5	1533	87.8	30.4
2000	2.09	1.424	1.98	0.99	6	3	0.75	550	0.05	24.8	1556	89.2	34.5

layer. Both types of bandgap grading deliver practically the same efficiency for $L_s \leq 500$ nm, but nonlinear grading is superior to linear grading for $L_s \geq 1000$ nm. Overall, *the nonlinearly graded bandgap delivers higher efficiency than the homogeneous and linearly graded bandgaps for all considered thicknesses of the n-AlGaAs layer.*

The optimal designs in Table 6.3 have $L_x = 525 \pm 25$ nm and $\zeta = 0.05$. The values of E_a, E_b, A, α, and ν are the same for all values of L_s; however, $K \in \{1, 2, 3\}$ does vary with L_s. The values of $A \sim 1$ and $E_b = 1.98$ eV are independent of L_s, the latter being significantly lower than its maximum allowed value.

Figure 6.7 presents the variations of E_g, χ, E_c, E_v, E_i, n, p, n_i, G, R, and N_f with z for the most efficient AlGaAs solar cell in Table 6.3. The magnitude of E_g is large near both faces of the n-AlGaAs photon-absorbing layer, which features elevate V_{oc} [4]. The regions in which E_g is small are of substantial thickness, these regions being responsible for elevating G [11]. The spatial profiles of E_c and E_i are similar to that of E_g. The intrinsic carrier density n_i is small where E_g is large and vice versa. The electron-hole-pair generation rate is higher in regions with lower bandgap and vice versa. The J_{dev}-V_{ext} characteristics of the solar cell indicate that $J_{dev} = 23.8$ mA cm^{-2} and $V_{ext} = 1450$ mV for best performance.

6.5   CIGS⊕CZTSSE SOLAR CELLS

Sections 6.2–6.4 provide examples of a two-terminal single-junction solar cell in which the bandgap energy of the sole photon-absorbing layer is nonlinearly graded. The coupled opto-electronic model of Chapters 2 and 3 [1] has also been applied for maximizing the efficiency of a two-terminal single-junction solar cell containing two photon-absorbing layers of different semiconductors. This format eliminates parasitic impedances and additional circuitry needed

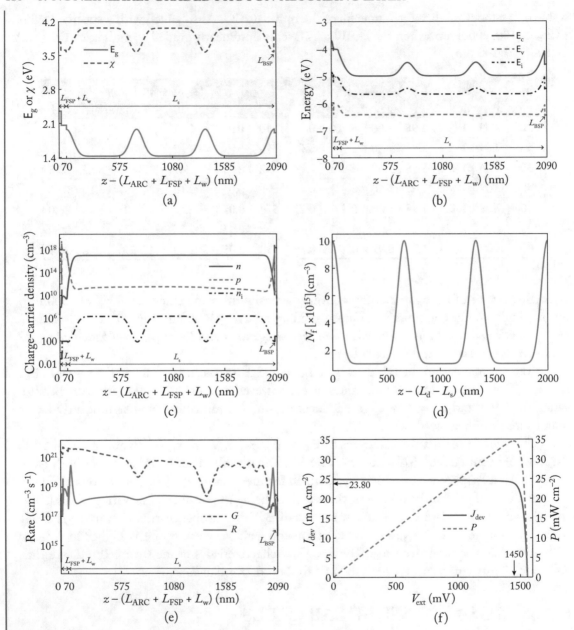

Figure 6.7: Variations of (a) E_g and χ; (b) E_c, E_v, and E_i; (c) n, p, and n_i; (d) N_f; and (e) G and R with z in the optimal AlGaAs solar cell with a 2000-nm-thick nonlinearly graded n-AlGaAs photon-absorbing layer. (f) Plots of J_{dev} and P vs. V_{ext}. Values of J_{dev} and V_{ext} for maximum P are identified. The bandgap and geometric parameters are available in Table 6.3.

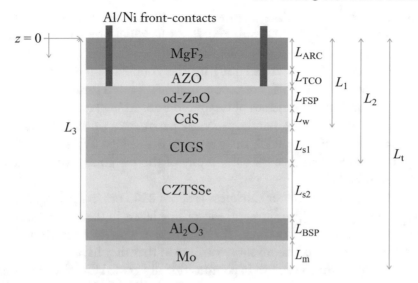

Figure 6.8: Schematic of a CIGS⊕CZTSSe solar cell with a flat back reflector.

for tandem solar cells [16]. However, the two photon-absorbing materials must have minimum lattice difference and should be capable of being deposited in a single device with compatible fabrication techniques. CIGS and CZTSSe are almost lattice matched [17, 18] and can be fabricated using vapor deposition techniques [19].

With Eq. (6.1) providing the template, the z-dependent bandgap energy (in eV) modeled in the CIGS layer of thickness L_{s1} as [20]

$$\mathsf{E}_g(z) = \mathsf{E}_{a1} + A_1 \left(\mathsf{E}_{b1} - \mathsf{E}_{a1}\right) \left(\frac{1}{2}\left\{\sin\left[2\pi\left(K_1\frac{z - L_1}{L_{s1}} - \nu_1\right)\right] + 1\right\}\right)^{\alpha_1},$$

$$z \in [L_1, L_2], \tag{6.2}$$

and in the CZTSSe layer of thickness L_{s2} as

$$\mathsf{E}_g(z) = \mathsf{E}_{a2} + A_2 \left(\mathsf{E}_{b2} - \mathsf{E}_{a2}\right) \left(\frac{1}{2}\left\{\sin\left[2\pi\left(K_2\frac{z - L_2}{L_{s2}} - \nu_2\right)\right] + 1\right\}\right)^{\alpha_2},$$

$$z \in [L_2, L_3], \tag{6.3}$$

where $\mathsf{E}_{b1} = 1.626$ eV, $\mathsf{E}_{b2} = 1.49$ eV, $L_1 = L_{ARC} + L_{TCO} + L_{FSP} + L_w$, $L_2 = L_1 + L_{s1}$, and $L_3 = L_2 + L_{s2}$. Whereas $\mathsf{E}_g(z)$ in the CIGS layer can be engineered through $\xi_1(z)$ [21], $\mathsf{E}_g(z)$ in the CZTSSe layer can be engineered through $\xi_2(z)$ [17, 22]. Available numerical results [2, 3] have amply demonstrated that a periodically corrugated back reflector is unnecessary and a flat back reflector is adequate. The 12-dimensional (i.e., $\tilde{N} = 12$) parameter space for maximizing the efficiency of the CIGS⊕CZTSSe solar cell is as follows: $\mathsf{E}_{a1} \in [0.947, 1.626]$ eV,

$A_1 \in [0, 1]$, $K_1 \in [0, 8]$, $\nu_1 \in [0, 1]$, $\alpha_1 \in [0, 8]$, $\mathsf{E}_{a2} \in [0.91, 1.49]$ eV, $A_2 \in [0, 1]$, $K_2 \in [0, 8]$, $\nu_2 \in [0, 1]$, $\alpha_2 \in [0, 8]$, $L_{s1} \in [0, 2200]$ nm, and $L_{s2} \in [0, 2200]$ nm, subject to the constraints $\mathsf{E}_{a1} \leq \mathsf{E}_{b1}$, $\mathsf{E}_{a2} \leq \mathsf{E}_{b2}$, $0 < L_{s1} + L_{s2} \leq 2200$ nm.

The highest value of η predicted [20] by the coupled optoelectronic model for the CIGS⊕CZTSSe solar cell is 34.45%; correspondingly, $J_{sc} = 38.11$ mA cm^{-2}, $V_{oc} = 1085$ mV, and FF = 0.83. The optimal thicknesses of the photon-absorbing layers are $L_{s1} = 300$ nm and $L_{s2} = 870$ nm. The corresponding bandgap-energy parameters are as follows: $\mathsf{E}_{a1} = 0.95$ eV, $A_1 = 0.91$, $K_1 = 1.88$, $\nu_1 = \nu_2 = 0.75$, $\alpha_1 = \alpha_2 = 6$, $\mathsf{E}_{a2} = 0.91$ eV, $A_2 = 0.99$, and $K_2 = 2$. The efficiency drops to no less than 34.43%, if any of the foregoing optimal bandgap-energy parameters is altered by 1%.

Whereas the standard absorber thickness is about 2200 nm in the single-absorber CIGS and CZTSSe solar cells [7, 23], both photon-absorbing layers in the optimal CIGS⊕CZTSSe solar cell are together only 1170-nm thick. The thickness of a photon-absorbing layer should be less than the diffusion length of the minority carriers so that they have a reasonable probability of being collected. That diffusion length in compositionally graded CIGS varies from 500–2000 nm [24], so that $L_{s1} = 300$ nm in the optimal CIGS⊕CZTSSe solar cell is definitely lower. However, the minority carriers generated in the CZTSSe layer also need to traverse the CIGS region to be collected across the p–n junction (i.e., the CdS/CIGS interface). As the minority-carrier diffusion length is around 1200 nm in CZTSSe [25], even the minority carriers generated deep inside the CZTSSe layer have diffusion lengths longer than or comparable to $L_{s1} + L_{s2} = 1170$ nm. Furthermore, the drift-field created by the E_g-gradient helps to accelerate the minority carriers toward the p-n junction [14].

If the CZTSSe layer is absent but $L_{s1} = 300$ nm, the highest efficiency predicted is 19.01% along with $J_{sc} = 25.98$ mA cm^{-2}, $V_{oc} = 1023$ mV, and FF = 73%. If the CIGS layer is absent but $L_{s2} = 870$ nm, the highest efficiency is 21.74% along with $J_{sc} = 37.39$ mA cm^{-2}, $V_{oc} = 772$ mV, and FF = 75%. Thus, the optimal CIGS⊕CZTSSe solar cell outperforms both CIGS and CZTSSe solar cells in all four performance parameters: η, J_{sc}, V_{oc}, and FF. The CIGS⊕CZTSSe solar cell appears to derive the high value of J_{sc} from the CZTSSe layer and the high value of V_{oc} from the CIGS layer. The CIGS⊕CZTSSe solar cell with both of its photon-absorbing layers graded nonlinearly also outperforms its analogs with linearly graded and homogeneous photon-absorbing layers [20].

The variations of E_g, χ, E_c, E_v, E_i, n, p, n_i, G, R, and N_f with z are presented in Fig. 6.9. The spatial profile of E_g in both photon-absorbing layers comprises constant-E_g regions separated by regions with large E_g gradients. The bandgap energy is low in the constant-E_g regions, these regions being responsible for elevating the electron-hole-pair generation rate [11], and G indeed exceeds R in these regions. The large E_g-gradient close to the back face of the CZTSSe layer enhances the drift field to reduce the back-surface recombination rate, thereby supplementing the role of the Al$_2$O$_3$ passivation layer [13, 26]. Since the bandgap energy is high close to both faces of each photon-absorbing layer, V_{oc} is high in the optimal design [12, 13, 27]. The

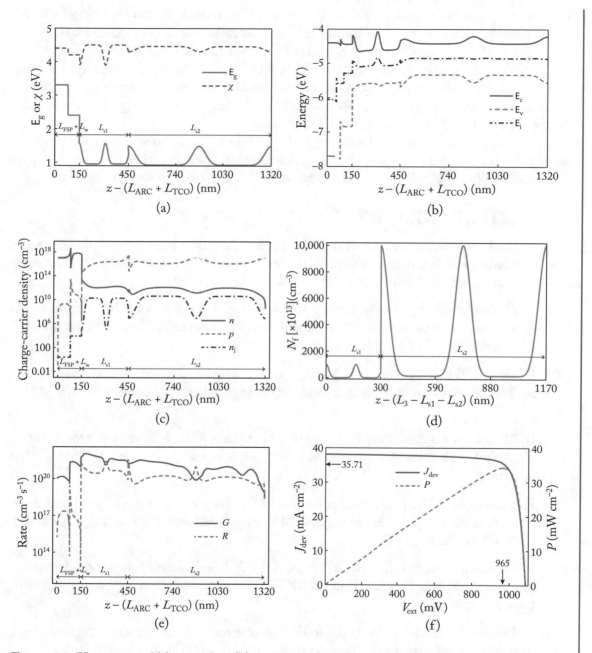

Figure 6.9: Variations of (a) E_g and χ; (b) E_c, E_v, and E_i; (c) n, p, and n_i; (d) N_f; and (e) G and R with z in the optimal CIGS⊕CZTSSe solar cell. (f) Plots of J_{dev} and P vs. V_{ext}. Values of J_{dev} and V_{ext} for maximum P are identified.

triangular regions in the middle of the spatial profile of E_g in each photon-absorbing layer also create an additional drift field that favors charge-carrier collection deep inside that layer [14].

The profiles of the minority charge-carrier density n and the majority charge-carrier density p in Fig. 6.9 are discontinuous at the CIGS/CZTSSe interface, the discontinuity of p being considerably larger than of n. These are consistent with the small discontinuity of the conduction band-edge energy E_c and the large discontinuity in valence band-edge energy E_v, which make E_g also discontinuous at the CIGS/CZTSSe interface.

The J_{dev}–V_{ext} characteristics shown in Fig. 6.9 indicate the solar cell should be operated with $V_{ext} = 965$ mV to deliver best performance with $J_{dev} = 35.71$ mA cm^{-2} so that $\eta = 34.45\%$. Since FF is high, $\eta = 33\%$ even when V_{ext} is reduced to 900 mV.

6.6   BIBLIOGRAPHY

[1] T. H. Anderson, B. J. Civiletti, P. B. Monk, and A. Lakhtakia, Coupled optoelectronic simulation and optimization of thin-film photovoltaic solar cells, *Journal of Computational Physics*, 407:109242 (2020). 99, 109

T. H. Anderson, B. J. Civiletti, P. B. Monk, and A. Lakhtakia, Coupled optoelectronic simulation and optimization of thin-film photovoltaic solar cells, *Journal of Computational Physics*, 418:109561 (2020) (corrigendum).

[2] F. Ahmad, T. H. Anderson, P. B. Monk, and A. Lakhtakia, Efficiency enhancement of ultrathin CIGS solar cells by optimal bandgap grading, *Applied Optics*, 58:6067–6078 (2019). 99, 111

F. Ahmad, T. H. Anderson, P. B. Monk, and A. Lakhtakia, Efficiency enhancement of ultrathin CIGS solar cells by optimal bandgap grading, *Applied Optics*, 59:2615 (2020) (erratum).

[3] F. Ahmad, A. Lakhtakia, T. H. Anderson, and P. B. Monk, Towards highly efficient thin-film solar cells with a graded-bandgap CZTSSe layer, *Journal of Physics: Energy*, 2:025004 (2020). 99, 103, 106, 111

F. Ahmad, A. Lakhtakia, T. H. Anderson, and P. B. Monk, Towards highly efficient thin-film solar cells with a graded-bandgap CZTSSe layer, *Journal of Physics: Energy*, 2:039501 (2020) (corrigendum).

[4] F. Ahmad, A. Lakhtakia, and P. B. Monk, Optoelectronic optimization of graded-bandgap thin-film AlGaAs solar cells, *Applied Optics*, 59:1018–1027 (2020). 99, 108, 109

[5] R. Storn and K. Price, Differential evolution—a simple and efficient heuristic for global optimization over continuous spaces, *Journal of Global Optimization*, 11:341–359 (1997). 99

[6] F. Ahmad, Optoelectronic modeling and optimization of graded-bandgap thin-film solar cells, Ph.D. Dissertation (The Pennsylvania State University, University Park, PA, 2020). 99

[7] P. Jackson, R. Wuerz, D. Hariskos, E. Lotter, W. Witte, and M. Powalla, Effects of heavy alkali elements in Cu(In,Ga)Se$_2$ solar cells with efficiencies up to 22.6%, *Physica Status Solidi RRL*, 10:583–586 (2016). 102, 112

[8] M. Gloeckler and J. R. Sites, Potential of submicrometer thickness Cu(In,Ga)Se$_2$ solar cells, *Journal of Applied Physics*, 98:103703 (2005). 102, 103

[9] J. Song, S. S. Li, C. H. Huang, O. D. Crisalle, and T. J. Anderson, Device modeling and simulation of the performance of Cu(In$_{1-x}$,Ga$_x$)Se$_2$ solar cells, *Solid-State Electronics*, 48:73–79 (2004). 102

[10] F. Ahmad, T. H. Anderson, P. B. Monk, and A. Lakhtakia, Optimization of light trapping in ultrathin nonhomogeneous CuIn$_{1-\xi}$Ga$_\xi$Se$_2$ solar cell backed by 1D periodically corrugated backreflector, *Proc. of SPIE*, 10731:107310L (2018). 102

[11] S. J. Fonash, *Solar Cell Device Physics*, 2nd ed. (Academic Press, Burlington, MA, 2010). 103, 106, 109, 112

[12] K.-J. Yang, D.-H. Son, S.-J. Sung, J.-H. Sim, Y.-I Kim, S.-N. Park, D.-H. Jeon, J. Kim, D.-K. Hwang, C. W. Jeon, D. Nam, H. Cheong, J.-K. Kang, and D.-H. Kim, A bandgap-graded CZTSSe solar cell with 12.3% efficiency, *Journal of Materials Chemistry A*, 4:10151–10158 (2016). 106, 112

[13] T. Dullweber, O. Lundberg, J. Malmström, M. Bodegård, L. Stolt, U. Rau, H. W. Schock, and J. H. Werner, Back surface band gap gradings in Cu(In,Ga)Se$_2$ solar cells, *Thin Solid Films*, 387:11–13 (2011). 106, 112

[14] J. A. Hutchby, High-efficiency graded band-gap Al$_x$Ga$_{1-x}$As–GaAs solar cell, *Applied Physics Letters*, 26:457–459 (1975). 106, 112, 114

[15] I. L. Repins, L. Mansfield, A. Kanevce, S. A. Jensen, D. Kuciauskas, S. Glynn, T. Barnes, W. Metzger, J. Burst, C.-S. Jiang, P. Dippo, S. Harvey, G. Teeter, C. Perkins, B. Egaas, A. Zakutayev, J.-H. Alsmeier, T. Lußky, L. Korte, R. G. Wilks, M. Bär, Y. Yan, S. Lany, P. Zawadzki, J.-S. Park, and S. Wei, Wild band edges: The role of bandgap grading and band-edge fluctuations in high-efficiency chalcogenide devices, *Proc. of 43rd IEEE Photovoltaics Specialists Conference (PVSC)*, pages 309–314, Portland, OR, June 5–10, 2016. 106

[16] M. A. Green, Photovoltaic technology and visions for the future, *Progress in Energy*, 1:013001 (2019). 111

[17] S. Adachi, *Earth-Abundant Materials for Solar Cells* (Wiley, Chichester, West Sussex, UK, 2015). 111

[18] T. Klinkert, M. Jubault, F. Donsanti, D. Lincot, and J.-F. Guillemoles, Differential in-depth characterization of co-evaporated Cu(In,Ga)Se$_2$ thin films for solar cell applications, *Thin Solid Films*, 558:47–53 (2014). 111

[19] K. L. Chopra, P. D. Paulson, and V. Dutta, Thin-film solar cells: An overview, *Progress in Photovoltaics: Research and Applications*, 12:69–92 (2004). 111

[20] F. Ahmad, A. Lakhtakia, and P. B. Monk, Double-absorber thin-film solar cell with 34% efficiency, *Applied Physics Letters*, 117:033901 (2020). 111, 112

[21] C. Frisk, C. Platzer-Björkman, J. Olsson, P. Szaniawski, J. T. Wätjen, V. Fjällström, P. Salomé, and M. Edoff, Optimizing Ga-profiles for highly efficient Cu(In,Ga)Se$_2$ thin film solar cells in simple and complex defect models, *Journal of Physics D: Applied Physics*, 47:485104 (2014). 111

[22] A. Kanevce, I. L. Repins, and S. H. Wei, Impact of bulk properties and local secondary phases on the Cu$_2$ZnSn(S,Se)$_4$ solar cells open-circuit voltage, *Solar Energy Materials and Solar Cells*, 133:119–125 (2015). 111

[23] W. Wang, M. T. Winkler, O. Gunawan, T. Gokmen, T. K. Todorov, Y. Zhu, and D. B. Mitzi, Device characteristics of CZTSSe thin-film solar cell with 12.6% efficiency, *Advanced Energy Materials*, 4:1301465 (2014). 112

[24] G. Brown, V. Faifer, A. Pudov, S. Anikeev, E. Bykov, M. Contreras, and J. Wu, Determination of the minority carrier diffusion length in compositionally graded Cu(In,Ga)Se$_2$ solar cells using electron beam induced current, *Applied Physics Letters*, 96:022104 (2010). 112

[25] T. Gokmen, O. Gunawan, and D. B. Mitzi, Minority carrier diffusion length extraction in Cu$_2$ZnSn(S,Se)$_4$ solar cells, *Journal of Applied Physics*, 114:114511 (2013). 112

[26] P. Casper, R. Hünig, G. Gomard, O. Kiowski, C. Reitz, U. Lemmer, M. Powalla, and M. Hetterich, Optoelectrical improvement of ultra-thin Cu(In,Ga)Se$_2$ solar cells through microstructured MgF$_2$ and Al$_2$O$_3$ back contact passivation layer, *Physica Status Solidi RRL*, 10:376–380 (2016). 112

[27] M. Gloeckler and J. R. Sites, Band-gap grading in Cu(In,Ga)Se$_2$ solar cells, *Journal of Physics and Chemistry of Solids*, 66:1891–1894 (2005). 112

Authors' Biographies

FAIZ AHMAD

Faiz Ahmad is a Lecturer in the Department of Physics at COMSATS University Islamabad, Pakistan. He received his M.Sc. (2010) and M.Phil. (2012) in Electronics from the Quaid-i-Azam University, Islamabad, Pakistan, and his Ph.D. (2020) in Engineering Science and Mechanics from The Pennsylvania State University, USA. His research interests relate to surface multiplasmonics and thin-film solar cells.

AKHLESH LAKHTAKIA

Akhlesh Lakhtakia is an Evan Pugh University Professor and the Charles Godfrey Binder (Endowed) Professor of Engineering Science and Mechanics at The Pennsylvania State University. He received his B.Tech. (1979) and D.Sc. (2006) in Electronics Engineering from the Institute of Technology, Banaras Hindu University, and his M.S. (1981) and Ph.D. (1983) in Electrical Engineering from The University of Utah. He has been elected a Fellow of the American Association for the Advancement of Sciences, American Physical Society, Institute of Physics (UK), Optical Society of America, SPIE–The International Society for Optics and Photonics, Institute of Electrical and Electronics Engineers, Royal Society of Chemistry, and Royal Society of Arts. He was the Editor-in-Chief of SPIE's online *Journal of Nanophotonics* from its inception in 2007 through 2013. His current research interests include: electromagnetic fields in complex mediums, sculptured thin films, mimumes, surface multiplasmonics, electromagnetic surface waves, solar cells, forensic science, engineered biomimicry, and biologically inspired design.

PETER B. MONK

Peter B. Monk is a Unidel Professor in the Department of Mathematical Sciences at the University of Delaware. He received his B.A. in Mathematics from Cambridge University, UK in 1978 and his Ph.D. in Mathematics from Rutgers University, NJ, USA in 1983. Since 1982 he has worked at the University of Delaware where he served as Chair of the department from 2007–2010. He is a Fellow of the Institute for Mathematics and its Applications in the UK. He is the author of *Finite Element Methods for Maxwell's Equations* (Oxford, 2003) and co-author with F. Cakoni and D. Colton of *The Linear Sampling Method in Inverse Electromagnetic Scattering* (SIAM, 2011). His current research interests are the modeling and optimization of thin-film solar cells, inverse electromagnetic scattering, and finite element methods.

Printed in the United States
by Baker & Taylor Publisher Services